MADE SIMPLE - MADE EASY

A selection of gadgets to assist the Model Engineer

by H. Maurice Turnbull

British Library Cataloguing-in-Publication-Data: a catalogue record of this book is held by the British Library.

ISBN No. 978-0-9564073-2-0

Published in Great Britain by:

Camden Miniature Steam Services
Barrow Farm, Rode, Frome, Somerset. BA11 6PS
www.camdenmin.co.uk

Layout and design by Andrew Luckhurst, Trowbridge, Wiltshire.

Printed by Imprint Design & Print, Newtown, Powys.

Contents

Preface

It is a privilege to write a preface to 'Made simple-Made easy'. The author is a practical mechanical engineer who exercised his skills in the exotic metal production industry, and at home studied the finer points found in nineteenth century machine tools and steam vehicles.

With a compact workshop in which model replica parts were made, it was vital that various simple devices, each to do a specific job, could be made to do other jobs without taking valuable workshop space. The devices, unlike the books of "101 useful gadgets", actually add to the precision of the model part.

As a certified mechanical engineer since 1942 and a member of the Society of Model & Experimental Engineers, I can vouch for the usefulness, the relative simplicity, cheapness and versatility of the items. The materials needed can often be found in the workshop off-cuts and recycling box. There is no need to drill holes in machine tools. Fitting is, as a policy reversible. Much of the work is unique.

Read the book to learn what is in it. Later, when a problem occurs, turn up the relevant chapter, read it carefully to gain help.

As a taster, look at chapter 4. This V-block is modelled after the commercial 'Keats' design of the early 20th century and copied since by the trade and amateur alike. Mr. Turnbull's design with a straight side, a strip of metal, and three nuts and bolts, can allow the throw of eccentrics to be located to within 0.0005, this repeatedly until a batch is complete, and quickly too.

Peter Spenlove-Spenlove DLC. BSc(hc)
Leicestershire.

Introduction

In this book is a collection of some of the gadgets that I have come up with after around forty years, off and on, in the world of model engineering. About half of them I have had published in various magazines, but of the rest I have not seen them anywhere in print before. The items I have chosen to include are the more simple ones, those that can be made up in a comparatively short time. I generally like to work on the principle that a gadget should not take much more time to make than the time it will save when in use. However I don't always keep to my principles as, in for example, the case of an attachment I made for the lathe. It took me three weeks to make, and fifteen minutes for it to cut the special worm wheel that I needed, and it has never been used since. This was a special case however, where accuracy was the criteria and time saving was of secondary importance. I have a few other similar examples, but they are not for this book.

Although obviously all the devices were made to be used in my workshop, they should all be easily adapted to other, different machines. It's the basic idea that is important.

The machines in question are,

1. A far Eastern Pinnacle PL1018 lathe,10in. swing by 18in. between centres, with a Norton type gear box.
Chapters 1, 2, 3, 5, 6, 7.

2. A Myford Super 7, with change wheels, and a quick change tool post.
Chapters,1, 2, 4, 5, 6.

3. A Boley-Leinen watchmaker's lathe.
Chapter 1.

4. A Centec 2B milling machine used with the vertical head.
Chapters 8 and 10.

5. A Taylor-Hobson engraving machine.
Chapter 13.

Acknowledgments

Special thanks to Sue Newbold and Ann Whorton for checking the draft grammatically and for presentation. To Neil Read for technical proof reading and John Slater for converting my dimensioned sketches into computer drawings. Peter Spenlove-Spenlove for the preface and all the other people who have encouraged me during the writing of the book.

PART ONE: DEVICES FOR THE LATHE

CHAPTER ONE FILING RESTS

There have been many filing rests described over the years but none I think with the basic simplicity of the one I wish to describe here. **Photo 1.1**. It was made for the production of small components such as nuts, bolts and fittings, and was made originally to fit in one of the six stations of a tailstock turret. **Photo 1.2**. There is no graduated movement of the rollers. Adjustment is achieved by placing packings of various thicknesses, under the roller slide, finer settings being achieved with shims or feeler gauges. The lathe is of far eastern origin called a Pinnacle 1018, i.e. 10 in. swing by 18 in. between centres. Setting up in the turret was very successful for production runs but was a little too fiddly for the occasional 'one off', so an adapter was quickly made which allows the rest to be popped into the four way tool post **Photo 1.3**. It can be seen from **Photo 1.4**, that the rest is small enough to clear the top slide, so can be left in position for as long as is required.

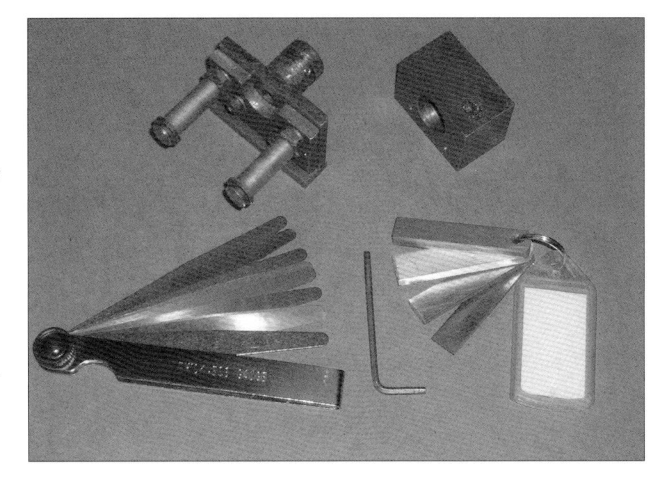

Photo 1.1

Dimensions, see **Fig 1.1**, are a guide only and can be altered to suit individual requirements. The only one which is important is the diameter of part A which is to fit in the turret, if that is what you want, otherwise suit yourselves. Measurements are Imperial, as many years ago when the rest was made, that was all we used.

Photo 1.2

The tool post mounting block is from 1 inch square and must be made on the lathe on which it is to be used. Fastened in the tool post, the hole 5/8 in.dia. is drilled, and preferably reamed from the chuck. The tapped hole for the grub screw can be located at an angle, as in my case, or in the end, it makes no difference. Part A is a simple turning job. The groove is to prevent any bruising by the clamping grub screw from interfering with the fit in the block or the turret. The 1/2in. spigot should be a loose fit in the hole of part B to allow silver solder to run in, perhaps about 0.005in. clearance. The back plate B is made according to the drawing, then silver soldered to A. If soldering is not possible then screw threads could be used but it would have to be a fine pitch as it is so short, British Standard Brass at 26 threads per inch would probably do. After fixing, chuck the round part and take a skim off the face of the back plate to make sure it is dead square to the axis.

Part C, the base plate is simple, the holes being counter-

Photo 1.3

Photo 1.4

bored or counter-sunk merely to keep the overall height as low as possible.

Part E, the roller slide is again a simple component, but here I would like to digress a little.

After I had made the rest it occurred to me that it would be an advantage if the top of the rollers were on the centre line of the lathe at their lowest position. This would make the setting to the correct height easier, and with the slide at its lowest position it would be perfect for making D-bits.

I made mine to the dimensions as shown in the drawing, but if I was to make another one I would do it in the following way. Machine all of E except the tapped holes for the rollers, then assemble all the parts already made and clamp in the lathe tool post. With a hardened point, gripped in the chuck, lightly scribe two lines using the saddle and cross feeds. Then dismantle and mark two further lines at a distance of half the diameter of the rollers below the first pair. Drill and tap the 2BA holes at the correct distance apart. This should put the top of the rollers on the centre height or if not there is no other way than to remove metal from the bottom of B or add shims between B and C as required. The notch in the top is merely to provide a little more clearance for long workpieces. Parts D, F, J and I need no explanation.

Rollers G were made from old high tensile cap headed screws, but mild steel would do. There has been some discussion in the past about hardening rollers and there are points for and against. I prefer not to heat treat. They should, however, be a nice running fit on H, without being sloppy.

Parts H are modified 0BA cap bolts. A length of 2.5in. should leave enough plain shank for the rollers to run on. Correct the length of thread if necessary then skim the under side of the head and slightly under-cut the shank. Machine away most of the head as in the drawing.

Assemble the parts, and your rest is ready to go to work. Slacken off the screws J and place packings equal to half the 'across flats' size required, under the roller slide. Tighten the screws and remove the packing. As can be seen in **Photo 1.1** I have a selection of packings on a key ring plus some others not shown, augmented with standard feeler gauges. As with a graduated type of rest it may be best to go a little higher to start with, and then lower the rollers after measuring the work, you can always take a little more off. An old turner once told me that even after a lifetime in the trade, he had still not come across a "putting on tool" !

And then - whilst I was writing this article I happened to spend some time looking at an old Super 7, newly acquired, and fitted with a quick-change tooling system. It may have been I was still subconsciously thinking of filing rests when it clicked in my mind that a Dickson type tool holder was well on the way to making another one. A length of square bar was quickly marked out from the lathe centre line, as already suggested, the holes drilled and tapped and the rollers transferred from the first rest. This is shown in **Photo 1.5**. Height adjustment would be made with the usual screw adjuster, which in my case has a pitch of 0.8mm, or .0315in. which is only 0.0003in. over 1/32in. If the bottom flange of the adjuster is marked with six divisions, then one division equals 0.1 mm or 0.0039in. (near enough 0.004in.).

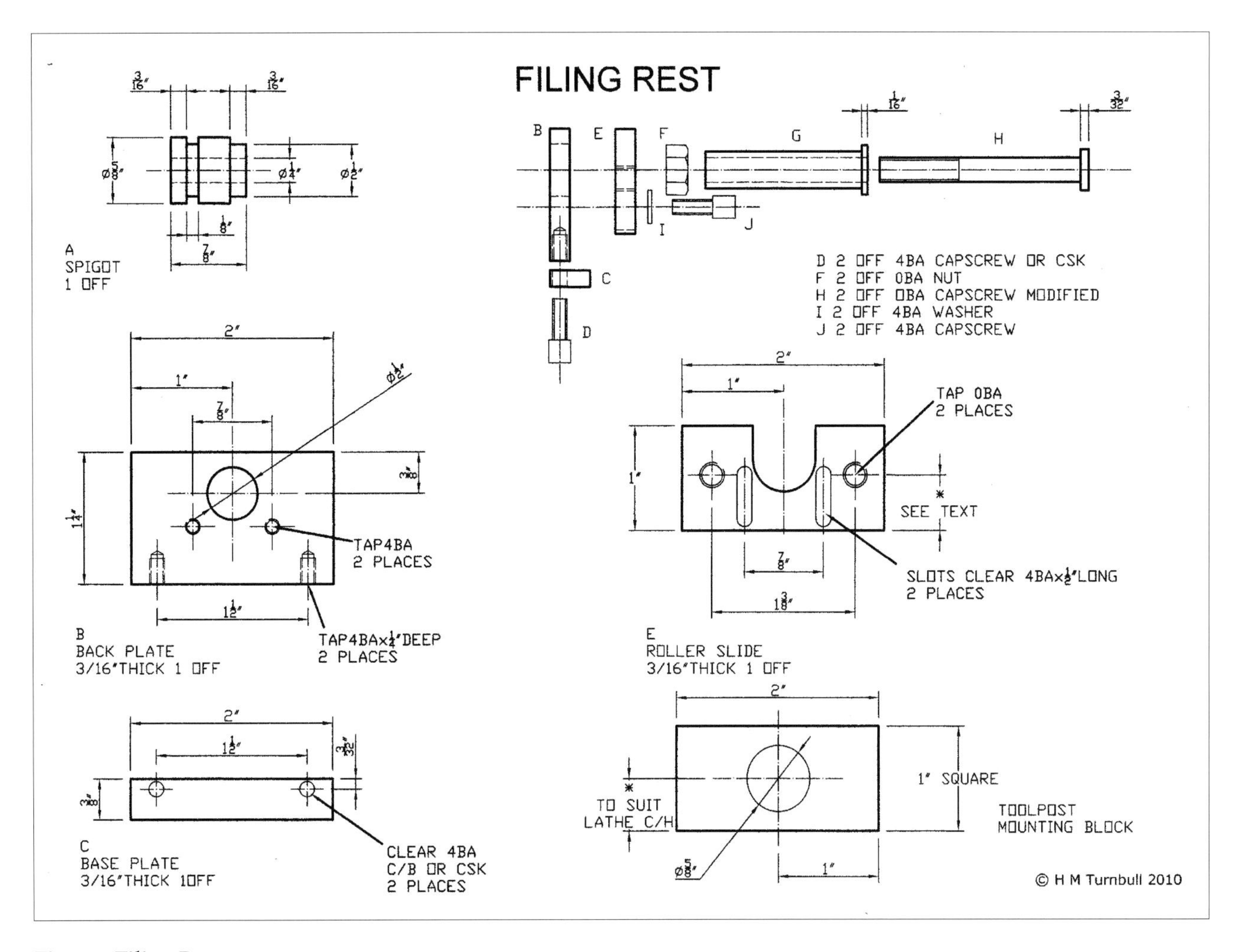

Fig 1.1: Filing Rest

The final item in this chapter will probably not apply to the majority of model engineers, but I have included it more for the idea than anything else. It involves the filing rest which is part of the equipment for a Boley-Leinen watch maker's lathe. As can be seen from **Photo 1.6** it is a beautifully made piece of equipment. The toothed wheel at the front meshes with a nut inside the central column that moves the cradle up or down. Unfortunately it does have a problem in that when in use, the vibrations from the file cause the wheel to rotate, so lowering the rollers. A watchmaker told me that this was a known fault. My solution is seen in **Photo 1.7**. A frame was made to encompass the central column and the toothed wheel, with a sprung detent in the side, which locates within the teeth and prevents rotation. **Photo 1.8** shows the rest mounted on the lathe. In keeping with another principle, of trying never to alter the original equipment, the filing rest has no modifications and the frame can be slipped on or off as required.

Photo 1.5

Photo 1.6

Photo 1.7

Photo 1.8

Chapter Two A STEADY SETTING DEVICE

This is a very easy tool to make. Its simplicity belies its usefulness because as will be discussed later, it can be used for several other jobs too.

For the tool as shown in **Photo 2.1**, first obtain your Morse taper shank. Don't overlook the possibility of using old taper shank drills when rummaging through the junk at car boot sales or searching the traders' stalls at exhibitions. A scrapped drill could still have a taper in good condition, and be had for a price which hardly makes it worthwhile getting the steel and switching the lathe on to make your own. You will usually find the shank past the end of the drill flutes is soft enough to machine, but beware of chuck arbors as these are hardened and no amount of heating on the one that I tried would soften it. If you want to make your own, then methods of doing this have been discussed in the model press many times. As a last resort you could buy a blank from the trade.

At the time I made this setter I was only using imperial measurements, but now I suppose I would use metric. However, I am going to stick to the old ones for the description.

Mount the taper in the lathe headstock and machine a location for the discs and for a length of thread. My spigot was 3/8in. dia. by 1/8in. long and the screw cut thread is 5/16 Whitworth by 3/8in. long. Whatever sizes you use make sure they are common and that you have a reamer and tap to suit. A nicely knurled brass nut completes the main parts.

The setting discs can be made from a variety of materials, iron, steel, aluminium alloy, brass, or even hard plastic. They are not going to get any wear except when being fitted to the shank. Make up a few from different diameters of stock, bring them all to a uniform thickness, say 1/4in. by machining both sides, then drill and ream the hole in the middle.

With the shank back in the headstock fit each blank in turn and machine them to size on the outside, choosing

Photo 2.1

Photo 2.2

Photo 2.3

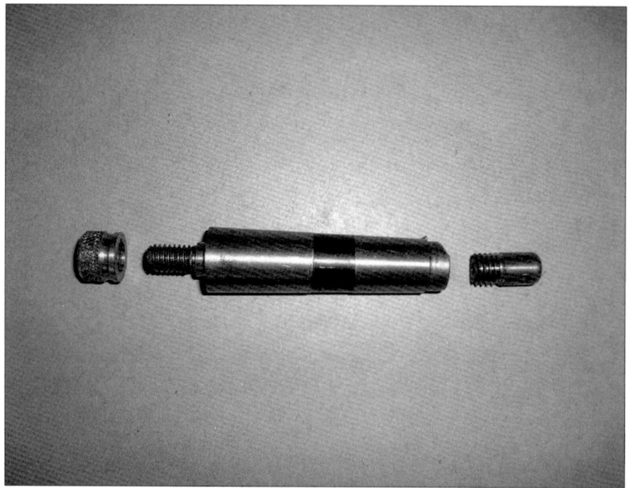

Photo 2.4

diameters that match the range of stock that you usually carry. Leave at least one blank un-machined to cover the odd size that crops up now and again. It takes only moments to machine it to size. **Photo 2.2** shows this operation on a 1in. disc and **Photo 2.3** is of the same one with the shank in the tailstock being used to set the fixed steady.

So that was the basic tool finished, but then I decided on the modifications shown in **Photo 2.4**, namely removing the tang, then drilling and tapping the end to take a draw bar. Now as well as having a setting tool I have a stub mandrel for second machining operations in the lathe and an arbor for use in the milling machine. A false tang is now needed, of course, to eject the shank from the tailstock.

SETTING A RADIUS.

Occasionally the need arises to machine an irregular shaped object to a precise diameter, such as the three lobed spacer shown in **Photo 2.5**. It cannot be measured by any normal means because it is not a full circle. If a disc of the required radius is used to set the lathe tool and the cross-slide dial set to, then any work-piece machined to the same mark must be correct! The theory has been tested. A setting disc was put on the shank in the tailstock barrel and the cutting tool was advanced to just touch a piece of shim resting on the edge of the disc. In this case it was 0.010in. brass. The cross-slide dial was then set to zero. After moving the saddle along the bed the tool was further advanced by the thickness of the shim and the dial again zeroed. The reason for using the shim is

Photo 2.5

Photo 2.6

Photo 2.7

Photo 2.8

for setting the tool with a sensitivity not possible using the lead-screw alone. Light cuts were then taken from the workpiece until the dial again read zero. As check, an accurately bored hole was made in a piece of scrap to serve as a ring gauge. When tried, the two parts fitted perfectly, no gap being visible at any position even when viewed through a strong magnifying glass. A further test with a spot of micrometer blue confirmed this. Photograph 2.6 shows an example.

For short, incomplete, internal radii on such things as saddle type boiler fittings the disc cannot be used to set the lathe tool, but they can be used as plug gauges to check dimensions as machining progresses.

Finally in **Photos 2.7** and **2.8** a disc is fitted into a vertical milling machine to locate an existing bore in a workpiece. The work shown is the steering arm from a full size farming machine, not exactly model engineering, but it is an example of the versatility of this simple gadget. This technique can also be applied to align an existing bore on the lathe face plate if a dial test indicator is not available.

The material used for the discs in the last four photographs is some hard plastic with a sheet of aluminium bonded to both sides. It had been dumped in a hedge bottom by some irresponsible person, but one man's junk is another mans treasure, and I now have enough material for dozens of discs.

CHAPTER THREE SUPER SLOW

In common with many model engineers, I have bemoaned the fact that the lowest speed on my lathe was not slow enough. The speed chart gives the slowest as 58 revolutions per minute, but it is actually 65 rpm. The lathe in question is my Pinnacle PL 1018. My intention was to cut some long, very small diameter, square threaded lead -screws, using my recently designed tool post milling machine. This required something like 140 slow turns of the mandrel. Yes I could have made a mandrel handle, and I could also certainly expect a very tired left arm. A smooth, slow, even rate of rotation is difficult to achieve with such a direct connection. A slow speed facility on the lathe would also be useful for coil or spring winding.

Some years ago in a magazine, a writer from New Zealand wrote an article about the problem of vibrations from the lathe motor mounted directly on the headstock. I had the same problem, so the motor has been removed from the lathe bed and mounted independently, it has also been fitted with a belt quick-release mechanism. I believe that, with the motor in the original position the modification I am about to describe would still work, although the changeover would be slower.

Basically, all I have done is to add another pair of pulleys to the drive from the motor.

Photo 3.1 shows the normal set up, the 3 driving peg holes in the V-belt pulley can be seen.

Photo 3.2 shows the arrangement of the new slow drive.

Photo 3.3 shows the back of the modified large pulley.

Photo 3.4 shows how the dishing clears the other parts of the headstock.

LARGE PULLEY

This is a 10.5in. poly-V pulley and is the component which gives trouble, -first of all you have to get one! Mine came from an old Hoover Electronic 1100 washing machine, released from the Domestic Engineering Department for recycling. Not everyone of course is going to have a

Photo 3.1

Photo 3.2

Photo 3.3

Photo 3.4

scrap front-loading washer to hand, so I went to a local market trader who sells domestic appliance parts and told him my story. He was very interested and gave me two suggestions: one (as a last resort) a new pulley could be bought as a spare part, or two, better still, companies specialising in washing machine repairs (see Yellow Pages) would probably be only too willing to part with an old one for the price of a pint. He also thought that Hotpoint machines used a similar type of pulley to the Hoover. Whatever the make it must be deeply dished, so that it clears the end of the mandrel and the gears.

So, you have managed to get one, now it has to be machined, and its too big for a 10in. swing lathe ! This is where a fellow club member, or a night school class, can be of help. The rim is set to run true and the back of the spokes just cleaned up for a diameter of about 3in. The centre boss with the slots for the locking tags is machined to the dimensions as shown in **Fig 3.1** and the original centre slot bored out. This completes the lathe work on the pulley. Make the drive plate as shown in **Fig 3.2**. Using the Plate as a jig, drill and tap the spokes 0BA, then fix with three countersunk screws. Next make the hub, **Fig 3.3**. The larger diameter is to be firmly fixed to the pulley so either make it a tight press fit or use adhesive. The small end is to locate easily into the bore of the lathe driven pulley. On my machine the end of the shaft for this pulley is 1/8in. short of the face, so the new hub in the large pulley should only protrude about 3/32in. from the face of the drive plate. Remove the V pulley from the lathe and temporarily fasten it to the large pulley. Jig drill three No.26 holes about 3/16in. deep and mark both pulleys for location. Tap the Drive Plate 2 BA. Make the three drive pegs, **Fig 3.4**, 2 BA cap screws are best as they are high tensile steel. Loctite into the drive plate.

Put a M6 x 30 cap headed screw through the hub, retain with an old O- ring or a piece of rubber tube pushed into the counter bore and the large pulley is finished.

SMALL PULLEY

The motor V pulley was a very tight fit and although keyways are cut in both pulley and shaft, no key was fitted. A grub screw in the bottom of the belt groove fits into the shaft keyway. The shaft which is sawn off flush with the pulley face, is tapped M8 and a bolt and a washer fitted. Why I don't know, because the pulley cannot possibly move axially. However this tapped hole is counter bored 8mm for a short distance and so provides a perfect centre location for the new poly-V pulley. Before making the small pulley fit the large one to the headstock and with a straight edge across the rim, check the distance to the face of the motor pulley. If yours is not 2.125in. then adjust this dimension to suit. Make the small pulley as shown in **Fig 3.5**.

The grooves are a direct copy of the washing machine motor shaft and although I used the standard (and cheaper) 6 rib belt I machined in 7 grooves to match the large pulley. Fit it into position on the motor and with a pistol drill, drill No.34 to a depth of about 3/8in. Mark for location, open up the flange to 4BA clear, tap the blind hole 4BA and refit. Deal with the other two holes in the same way, then fasten with three cap-head screws and leave in position permanently.

The belt is a standard one of 6 ribs - J Section. The length, which is the circumference measured in millimetres can only be determined by reference your own set-up.

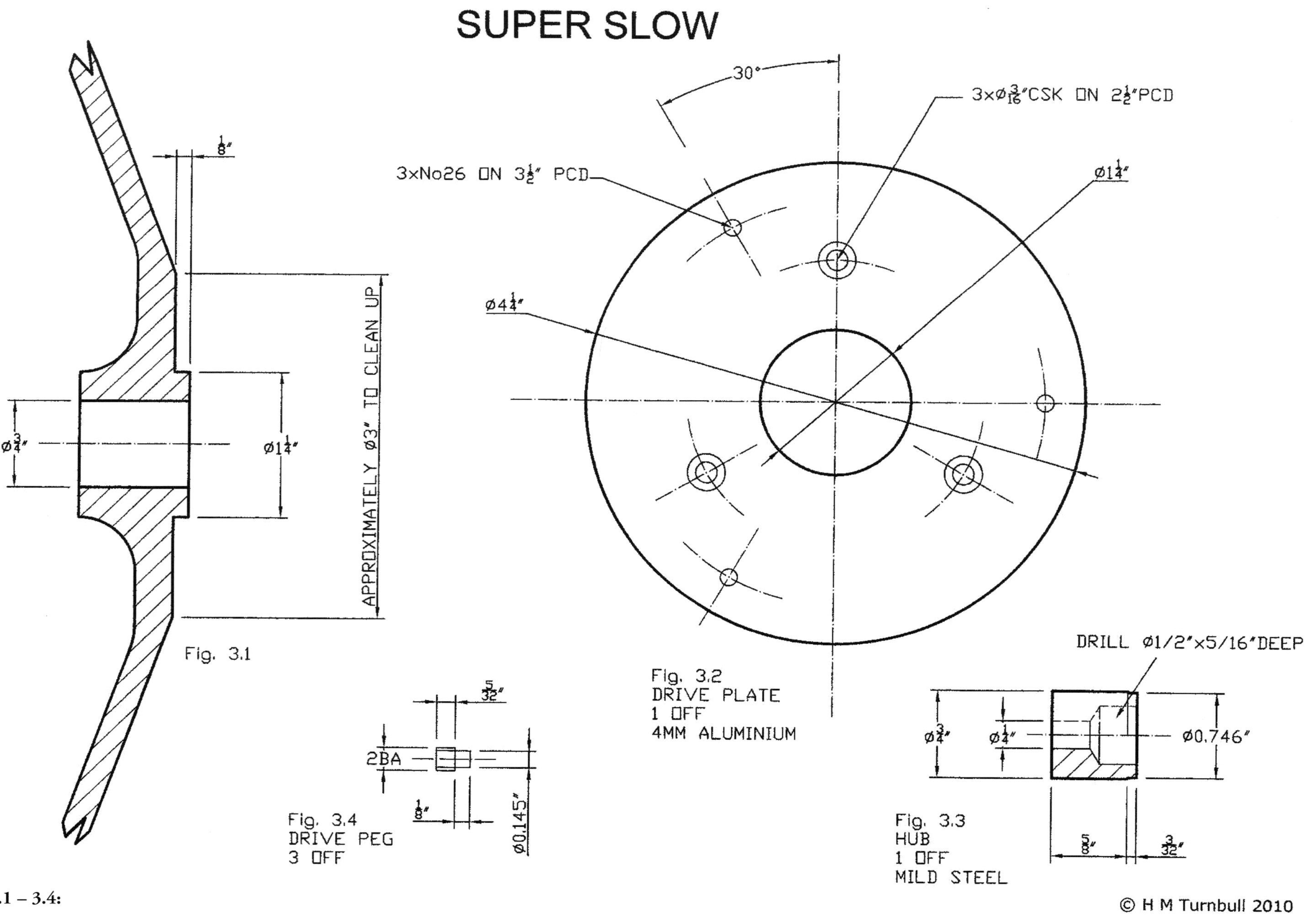
SUPER SLOW
30°
3×Ø3/16"CSK ON 2½"PCD
3×No26 ON 3½" PCD
Ø1¼"
Ø4¼"
1/8"
Ø3/4"
Ø1¼"
APPROXIMATELY Ø3" TO CLEAN UP
Fig. 3.1
Fig. 3.2
DRIVE PLATE
1 OFF
4MM ALUMINIUM
DRILL Ø1/2"×5/16"DEEP
Ø3/4"
Ø1/4"
Ø0.746"
Fig. 3.3
HUB
1 OFF
MILD STEEL
5/8"
3/32"
5/32"
2BA
1/8"
Ø0.145"
Fig. 3.4
DRIVE PEG
3 OFF
© H M Turnbull 2010

Figs 3.1 – 3.4:

SUPER SLOW

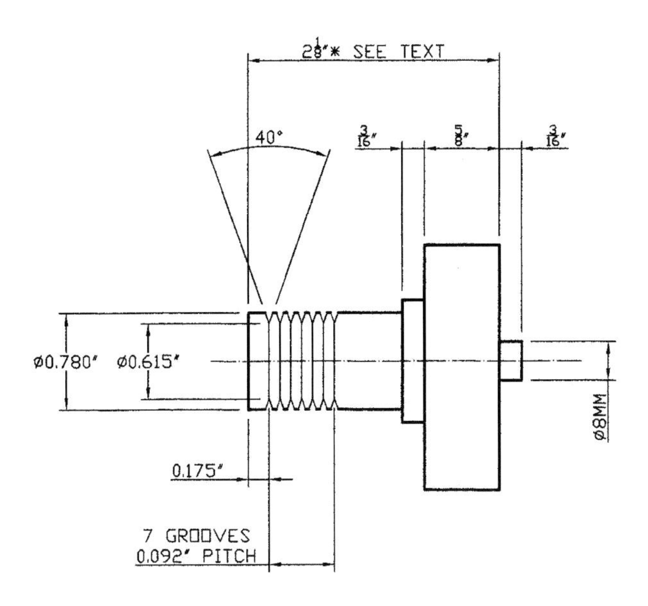

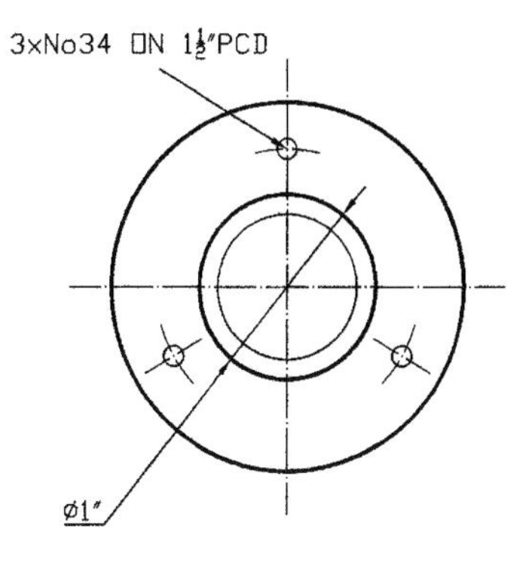

Fig 3.5
SMALL PULLEY
1 OFF ALUMINIUM

Fig 3.5

The system is now complete. Excluding the removal of the guards, the two systems can be changed in 40-50 seconds. Here's how. Release the motor tension, remove the V-belt, remove the retaining bolt from the headstock pulley, fit the large pulley and tighten the central screw, place the poly-V belt over both pulleys, re-tension the belt by moving the motor.

With the gears built into the headstock three new lower ratios are now available, by direct counting these are 38, 25 and 10.5 revolutions per minute.

If you wish to know the speed of a rotating shaft at something about 80 rpm or less, don't try counting a revolving mark whilst at the same time trying to look at a watch. Use one of those kitchen timers that count down and then sound a buzzer. Your eyes do the counting, your ears do the timing, it is easy.

A guard has yet to be made to cover the new drive, but when it is it will be a substantial wooden one. Geared head lathes do tend to make a noise and I have found that the cast alloy guard supplied with the Pinnacle magnifies it to an uncomfortable level. I am not only thinking of my own comfort but, because my workshop is in the cellar of a small terraced house, I have to consider my neighbours and try to reduce noise as much as possible.

At some time in the future, I intend to make a handle that will bolt onto one of the spokes of the large pulley so that I can turn the mandrel by hand if I want to, then for a steady input speed I can get out different chuck speeds, with an instant stop.

On ending this chapter, may I just give a word of warning !

The large pulley covers the end of the hollow mandrel, so don't put anything into the chuck that will protrude beyond the headstock or you will do some damage.

Chapter Four A FABRICATED ANGLE PLATE

This was the first article I ever had published, and that was back in 1991.

The first angle plate was made to be used on my old treadle operated Henry Milnes lathe (c1890) but it did not need any modification to make it suitable for a Myford S7. The photographs show a second one that was made to illustrate construction. It was made to metric dimensions.

Photo 4.1 shows it mounted on the face-plate to make an eccentric. The stock was made to run true and then a piece of flat bar was bolted to the face plate in contact with the right hand side of the angle plate, which has been machined to a straight edge. The lower edge of the plate has also been machined straight and at exactly right angles to the side. A round block is bolted below the face-plate at a distance equal to the offset required for the eccentric. This is set using any suitable form of packing. The outside of the eccentric was then machined.

Next, the mounting bolts were loosened and the angle plate slid down until it touched the round block, keeping it hard against the side bar, before re-tightening, and the hole for the shaft machined. The angle plate can then be returned to its original position by using the same packing as was used to set the block in the first place and the eccentric parted off. This can be repeated for as many times as required, and all the eccentrics will be the same.

CONSTRUCTION

The back is from 10 x 115 x 130mm plate, the angle from 50 x 50 x 6mm, cut down to 40 x 40 x 6mm. The U-clamp from 8mm round and the bridge is from 30 x 12 flat, 95mm long.

If necessary, alter any of the dimensions to suit your needs.

Photo 4.1

Photo 4.2

Photo 4.3

First, obtain your piece of angle iron, slightly longer than required, in my case about 95mm long, and make up the simple U-bolt and bridge piece. Putting a chamfer on the legs of the angle makes for a neater fit. Next put any old piece of round, (within the capacity of the angle) in the lathe, protruding a little more than the length of the angle, support with the tailstock and take a light skim off the diameter to true it up. Clamp the piece of angle onto this mandrel, over-hanging the end, using its own U-bolt, and face the end off, **Photo 4.2**. This ensures that the end is square to the inside of the V. Repeat for the other end. From a suitable piece of mild steel make up the back plate. One side and the bottom edge should be filed or milled straight and square to each other. Drill and slot to face-plate requirements. **Photo 4.3** shows all the parts. The back plate and the angle can now be held together with a large C-clamp, **Photo 4.4** and welded on the outside of the angle only.

Photo 4.4

To finish off, once again chuck a piece of round in the lathe, this time slightly shorter than the length of the angle and skim true on the diameter. Clamp the angle plate to it as shown in **Photo 4.5** and face off to make sure that any distortion from the welding is removed and that the rear of the bolting face is true to the inside of the angle.

Photo 4.5

Chapter Five SHORT TAPERS

I had a need to make several components with a Jacobs Taper No.1 on them. Some years ago I spent a very long time making a JT No.0 taper by trial and error. Fortunately the shaft it was on was long enough for me to make several attempts at it. This would not be possible with the proposed new parts, which would have to be right first time. I have often tried to think of a simple way of accurately setting the top slide over, from a device held between centres, but have never got anywhere with it. I know it can be done if you have an existing taper to copy, but what if you haven't?

I am a compulsive collector of tools and machines, as far as my budget will allow, so once in a moment of weakness, I bought a cheap sine bar, just for the sake of having one. I read in a magazine a description of a taper turning attachment for a Myford, which is basically a long sine bar. A Eureka moment later I had worked out on a scrap of paper how I could use my small (5in.) sine bar to set the top slide over with great accuracy. The resulting device is shown in **Photo 5.1**. A sort through the 'come in handy' box provided enough materials to start on it straight away.

It was made for use on the Pinnacle PL 1018 , but as can be seen from the photograph, the simplicity of the design should allow it to be easily adapted to other machines, but more of that later.

The platform fits over the cross slide and provides a datum face exactly aligned with the lathe centre-line. Two ledges, the long one of which is adjustable, carry the sine bar and the packings to give the required angle. The loosened top slide can then be swivelled against the bar before being tightened down.

The device as shown in the photographs is for a male taper. For female tapers, swap the ledges round and turn the sine bar over. If no packings are used then the top slide can be set truly parallel to the lathe centre line.

Photo 5.1

There are not many parts to it, and all are simple to make. They are;

Fig 5.1 - one platform.

Fig 5.2 - two supports

Fig 5.3 - one datum bar.

Fig 5.4 - one long ledge, post and knob.

Fig 5.5 - one short ledge.

Fig 5.6 - two finger screws.

various screws and dowels, and - one five inch (125mm) sine bar, slightly modified. The type with the rollers

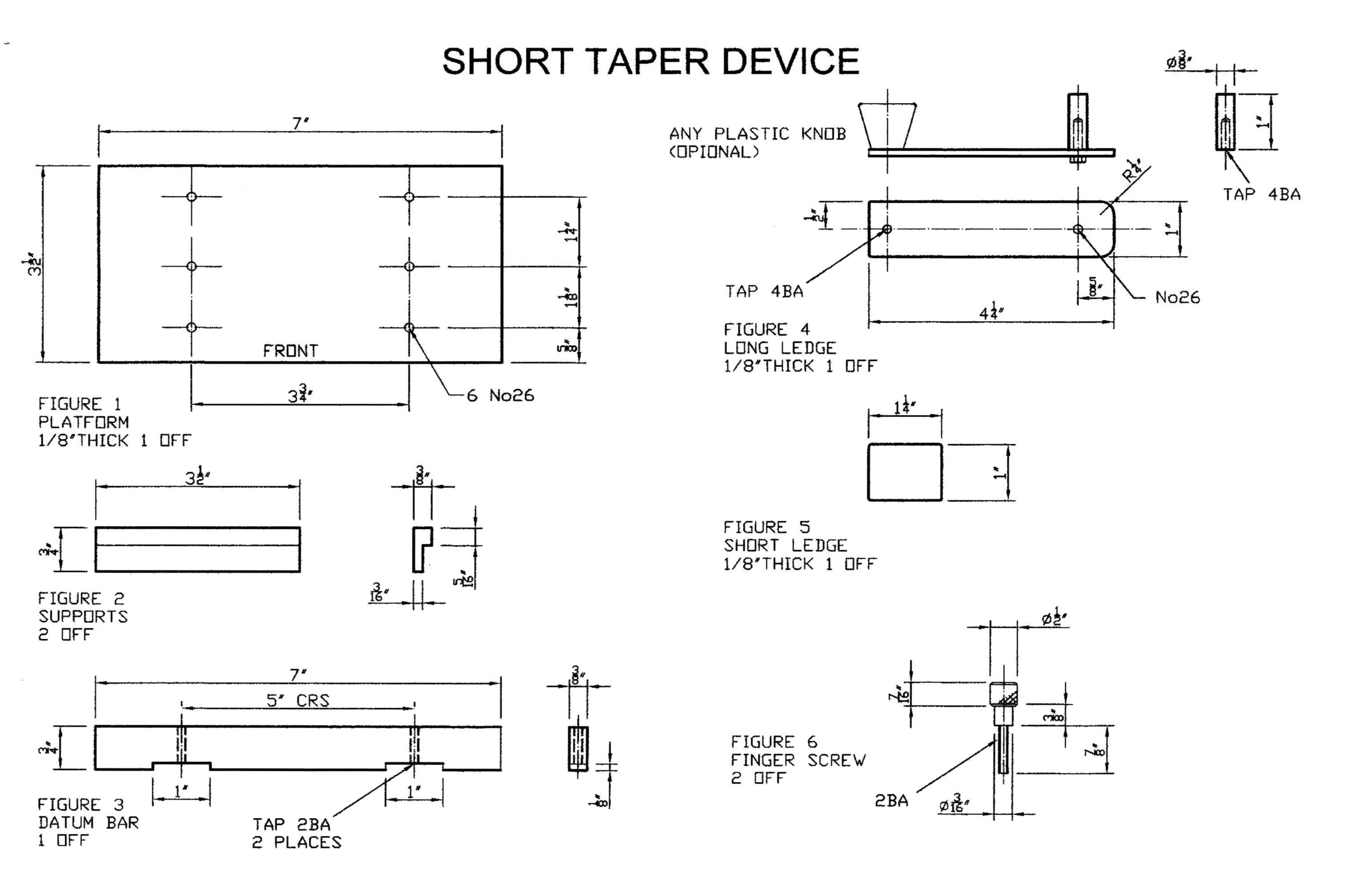
SHORT TAPER DEVICE
7"
3½"
1¼"
1⅛"
⅝"
FRONT
3¾"
6 No26
FIGURE 1
PLATFORM
1/8"THICK 1 OFF
3½"
¾"
⅜"
5/16"
3/16"
FIGURE 2
SUPPORTS
2 OFF
7"
5" CRS
¾"
1"
1"
⅜"
⅛"
FIGURE 3
DATUM BAR
1 OFF
TAP 2BA
2 PLACES
ANY PLASTIC KNOB
(OPIONAL)
ø⅜"
1"
TAP 4BA
R¼"
½"
1"
TAP 4BA
⅝"
No26
4¼"
FIGURE 4
LONG LEDGE
1/8"THICK 1 OFF
1¼"
1"
FIGURE 5
SHORT LEDGE
1/8"THICK 1 OFF
ø½"
7/16"
⅜"
⅞"
FIGURE 6
FINGER SCREW
2 OFF
2BA
ø3/16"

Figs 5.1 – 5.6

Photo 5.2

Photo 5.3

sitting in V-grooves and secured with the screws through the face was used.

The following notes refer to a device to be used on my machine as already stated and the dimensions will be Imperial because of the materials used. All can be changed to suit, but any that need special comment, marked by an * , will be explained.

1. Fig 5.1 The platform is made from aluminium plate but any material will do as long as it is flat. The centres of the two rows of holes (*) must be the same width as the cross slide on which it to be used, they are clearance for 4 BA.

2. Fig 5.2 Two supports. These were made from brass for two reasons. One, because I had some and two, I thought there would be a problem with distortion if using bright mild steel, and only milling one side away. Hot rolled mild steel might be all right. Do not drill any mounting holes at the moment. This (*) dimension is to let the platform and sine bar clear the top slide flange and will need adjusting to suit the lathe.

Set one of the supports square to the front edge of the platform and central under one of the rows of holes and clamp in place. Spot through, then drill and tap the three holes 4 BA. Fix with screws. A couple of dowels can be inserted for extra security. Making sure that the cross slide is clean, position the platform with the fixed support resting on top, then place the other one on the other side and put a clamp across them both. One of those light bar type clamps is ideal. Spot through the platform with a hand drill, and again drill and tap 4BA, and fix with screws. The platform should now be a firm sliding fit on the cross slide. Slightly enlarge or elongate the holes if necessary but there must be no shake whatsoever. Photograph 5.2.

3. Fig 5.3 The datum bar is made from the same material as the supports and for the same reason. The recesses for the ledges (*) are a nominal 1in. x 1/8in. just enough for them to slide through without too much slop. Another point is to make sure that the face for the sine bar is exactly square to the base.

4. Clamp the datum bar to the front edge of the platform then drill and fit a dowel right at the end that will be nearest to the chuck. Lock the lathe spindle and put into the chuck a rod that will hold a dial test indicator sticking out about 7 inches.

5. Lightly clamp the free end of the bar with a tool maker's clamp and fit the whole lot to the cross slide. By running the saddle up and down the bed, readings can now be taken at both ends of the datum bar. **Photo 5.3**. Adjust until both readings are the same. Clamp up tightly and

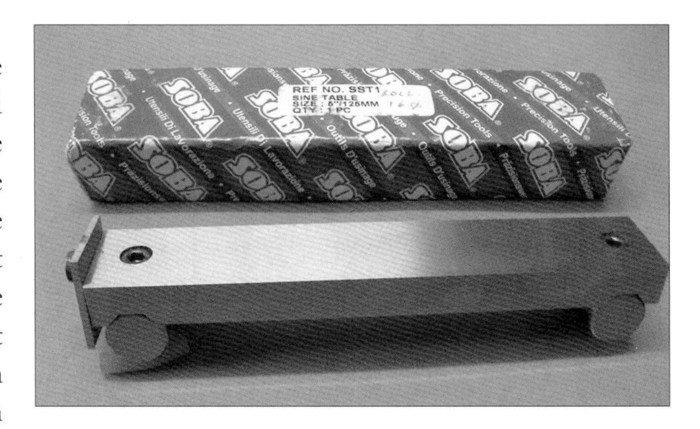

Photo 5.4

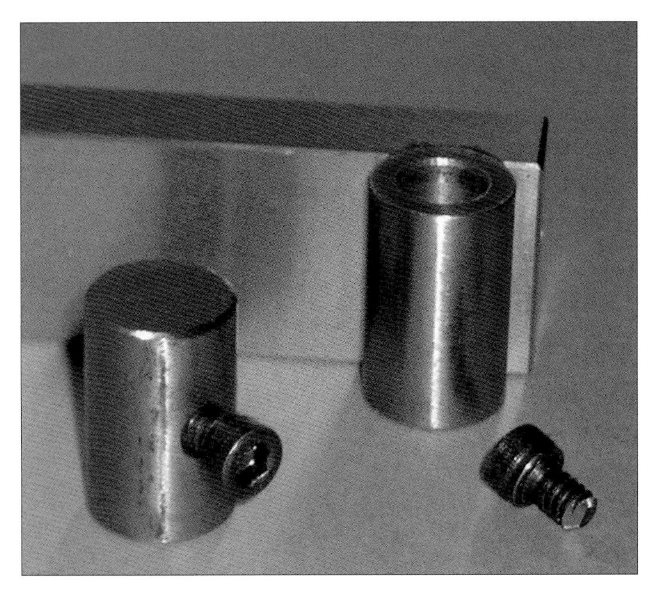

Photo 5.5

Photo 5.6

fit a second dowel. Drill and tap 4BA in about four places underneath to fix the bar down. The bar is now exactly parallel to the lathe centres though well below them. My datum bar had a dip of .003in. in the middle but was exactly the same over the whole area where the sine bar rollers rest, so this was acceptable.

6. Fig 5.4 The long ledge is made to the dimensions shown.

The post must be a good fit in the Sine Bar roller, with no slop.

7. Fig 5.5 The short ledge is a 1 1/4in. length of 1in. x 1/8in. bright mild steel flat with the corners rounded off.

8. Fig 5.6 Two finger screws are made as shown in the drawing.

9. The Sine bar needs some slight modification, but before starting I checked it for accuracy as I have found that some of these imports are not all that they should be. First the diameters of the two rollers were checked and found to be identical as were the distances from the top of the rollers to the face of the bar, however the roller centres were not correct. The size is stated as being 5in./125 mm!! and it was neither, but I suppose that as long as the true distance is known and used in the calculations then it will be all right. **Photo 5.4**. One of the rollers requires a 3/8in. reamed hole through it and rather than adapt an existing one I made a new one, and secured it with a 2 BA screw instead of the original 3/16 Whitworth screw. That is all that is required. The end fence is not used. **Photo 5.5**. From the photograph it is obvious how the whole lot goes together.

At the beginning I said its first intended use was to make JT1 drill chuck tapers. I am very rusty on trigonometry but eventually after a lot of head scratching I calculated that the packing needed to be .1933in. for such an angle. Because I don't have any slip gauges (yet), I decided to turn a piece of round to the said diameter and use that to set the Sine Bar. **Photo 5.6**. The device was then fitted to the cross slide and the top slide set before removing it. **Photo 5.7**. On a suitably prepared piece of round a test cut was made which, in my eagerness, was too deep, so the taper turned out to be too long and had to be faced off to the correct length. A trial fit with a new drill chuck, especially bought for the experiment, did not show any signs of binding at the top or bottom, so the round was removed from the lathe and the drill chuck firmly seated with a rubber mallet. It was tight, very tight, so I officially announced, to myself, that the device was a success. **Photo 5.8**.

There are a few points to watch for if the device is to be made for other lathes.

The cross slide must have a machined top and edges and the top slide have some part of that side that faces the device finished in the same way. I have examined four Myford Super 7s of various vintages and found that they have a reference ground on them. My own Super7, circa

Photo 5.7

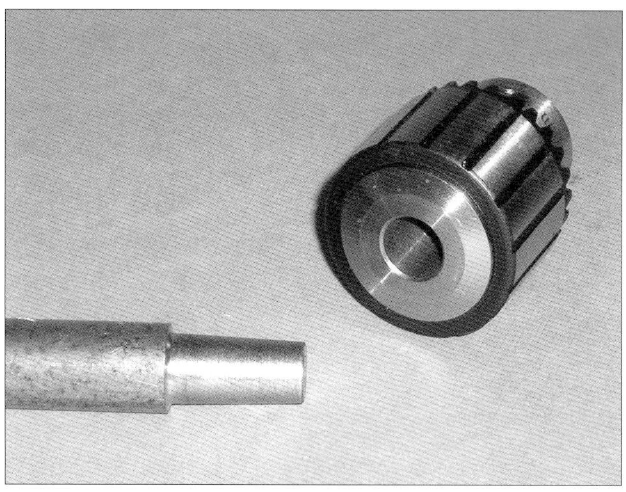

Photo 5.8

1960 has one but it had been painted over. It has to be assumed that this reference is truly parallel to the V-slides, but I think that it is a safe assumption.

I am fortunate that on the PL 1018 all the surfaces of the cross and top slides have a ground finish and that the cross gib strip is of the long taper type which is adjusted from the end, so that there are no screws in the side to get in the way.

These instructions and a sine bar were given to a friend of mine for him to test out for me by making the device for his Super 7. It was just as well that I did as it threw up a few difficulties. The cross slide on the Super7 is slightly wider than the Pinnacle, so the cap screws securing the supports are over where the ledges need to be.

There are however several possible solutions.

First, the screws themselves could be changed for countersunk ones.

This is all right for fixing the first support but could cause problems if the second support has to be adjusted to get exactly the right fit on the cross slide and of course both sides have to be the same because of the need to swap the ledges over for internal or external tapers. Alternatively, the holes in the supports could be moved in so that they are over the step in the support instead of down the centre. This would give a little more clearance over the outside.

Secondly, the ledges need not be 1" wide. I think that 3/4" wide would still be all right with the datum bar slots made to suit.

Thirdly, although it would be best to maintain the 5"centres of the slots, they need not be in the centre of the datum bar if off setting them a little to one side would help.

If you want to make the device, then please check your machine and decide if you need to use any of the above alternatives before you start cutting metal.

AFTERTHOUGHTS

As I said before, I don't have any slip gauges so for the test I turned a packing to .1933in. diameter. Afterwards, for an entirely different reason I was looking up the charts listing British Standard Sizes of Superseding Obsolete Drill Gauge and Letter Sizes, in the Model Engineers' Handbook, (2nd. Edition) and it struck me that there we have in the drill shanks a big range of packings. There are 80 in the number drills alone, usually with a variation of between 0.001 to 0.005in. and a whole alphabet in Letter drills. Check with a micrometer as some of the shanks are very slightly smaller than the flutes. Throw in the metric range of drills and a set of feeler gauges and the number is enormous. A number 10 drill is 0.1935in. only 0.0002in. away from the size I required for my drill chuck taper.

Like most lathes found in the amateur's workshop, my top slide does not have enough travel to make the most common number 2 Morse taper. but why do we need full length tapers? Industry uses short MT's, so why can't we? The PL 1018 has a 4MT in the mandrel, but the 4-2 converter is only 2in. long and is perfectly all right. If it is a necessity that the taper should be self ejecting, then machine a parallel tail on it, or drill and fit an extension to bring it to full length.

CHAPTER SIX OFFSET MEASURING

Some times it is necessary to locate a tool, such as a drill or trepanning tool at a specific distance from the lathe centre line, even if a reference point does not exist. Suppose a flange blank needs the holes to be drilled and there is no centre to measure the pitch diameter from, then this simple piece of equipment comes into its own. It is merely a holder that holds a rule aligned with the lathe centres.

Two versions are shown. The first with two columns, **Photo 6.1**, was made for the Pinnacle (PL1018), many years ago and has been very useful since. The second one, to fit the Myford, was made especially for this book. No dimensions or drawings are given because they can be made in a variety of ways to suit any machine, using odd bits of materials that may be to hand. Mine were made from blocks of aluminium alloy which I cast now and again in my own facilities, but any odd off-cuts of steel or iron could be used.

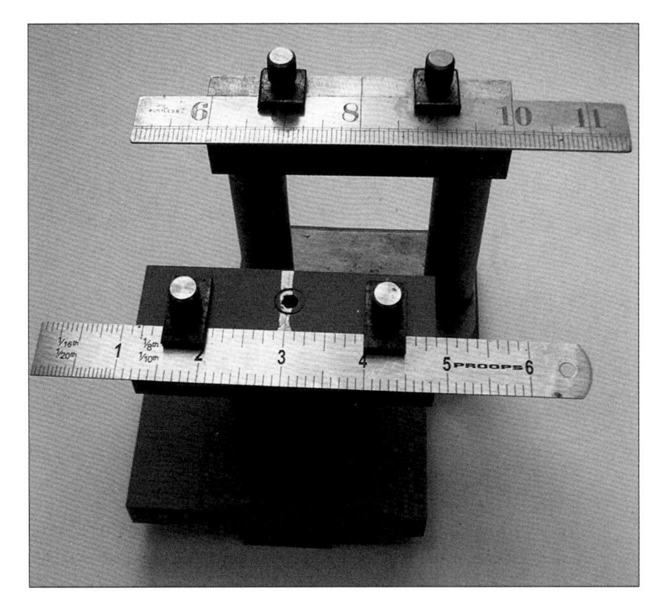

Photo 6.1

The first part to consider is the base plate which is to sit flat on the lathe bed. This must be able to be accurately and repeatedly located, on the PL by a groove that fits over the V-slide on the rear of the bed, and on the Myford by a block that sits between the shears. Next the column or columns are fastened to the bases with countersunk or counter-bored screws. The horizontal rule bar needs to be finished with the top on the lathe centre line, so I would suggest that the following procedure is used.

Fasten an over thick bar on top of the column/s then rest the device in position on the lathe bed against a sharp pointed centre in the tailstock. A slight tap with a soft hammer will leave an indentation in the bar at the centre height. The bar can then be milled down to this mark, however before doing this, I would suggest that the rule to be used is chosen. This can be just an ordinary 6in. rule as used in the workshop but it would be a great advantage if it is one where the graduations go all the way across from top to bottom. The reason will be explained later. Mill the bar to the indentation then without removing it from the vice, mill a step the depth of which is a little less than the thickness of the rule and about three quarters of the width. Remount it on top of the column/s making sure all is square. Two small clips, tightened with finger nuts, keep the rule in place.

Finally we come to the calibration. Place the gadget on the lathe with the rule lightly clamped. Again with the sharp pointed centre in the tailstock, align a graduation on the rule with the point, using a strong magnifying glass helps, and tighten up the rule clips. With a square, scribe a fiducial line from the rule across the top of the horizontal bar. The rule can then be removed for ordinary work and easily replaced to the fiducial line when needed.

I chose a trader's promotional rule that was marked with Imperial on one side and metric on the other. The

Photo 6.2

Photo 6.3

Imperial side is marked with sixteenths and eighths of an inch on one edge and twentieths and tenths on the other. The metric, with millimetres on one side and half millimetres on the other. Any of the four sides can be aligned with the fiducial mark.

As an alternative, a rule that is worn out on the end and so is no good for general work can be used. Only the middle graduations are needed, and if necessary it can be shortened to a convenient length.

Photo 6.2 shows the Myford device used for setting a trepanning tool, and **Photo 6.3** shows the PL1018 set up for drilling holes in a flange.

As an afterthought, if the Myford device is placed the wrong way round on the bed, the top of the rule carrier can act as a tool centre height setting device.

A critic might say, "Why not measure from a centre in the tailstock with a rule by hand?" Yes, but there are limitations. As an example, if using the tool post drilling spindle, with the tailstock pushed right up to the saddle and the barrel wound right out, the centre point will still not be far enough forward to allow measurements. Secondly, on any machine, the smallest radius that can be measured is equal to half the diameter of the tailstock barrel, and very much more than that if a tool holder or drilling spindle have to be taken into account. Both these devices allow measurements right up to the centre line and even beyond, even towards the back of the machine. "Why would anyone want to do that"?, might be asked. Read the next chapter.

Chapter Seven A TOOL POST DRILLING SPINDLE

There is nothing unconventional about this drilling spindle, it was designed only for very light drilling so there is no attempt at preloading and end thrust is taken by the ball races. The chuck capacity is 1/4in. but the motor is not strong enough to drive such a drill. If holes of this size and bigger are needed then they are positioned with a small centre drill and then opened out with the bench drilling machine. There are many designs available for spindles including those in a book in the Workshop Series, Number 27 by Harprit Sandhu showing a whole range of them. My claim for novelty in this design lies in the mounting. Mine was from a home made aluminium casting but a suitable sized block of any metal would do. The whole set up can be seen in **Photo 7.1**. It will be noted that the base is not machined flat as would be expected but has a step on one side positioned under the spindle **Photo 7.2**. My lathe has a ground finish on all faces of the top slide and the tool post stud is exactly the same distance from the edge as from the end. This means that the spindle, using the step on the mounting, can be easily located exactly parallel to the lathe centre line or exactly at right angles to it, assuming that the top slide is correctly set.

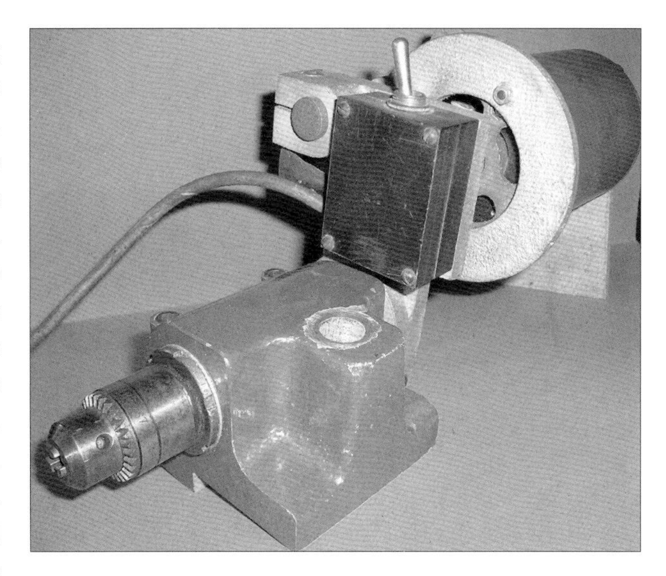

Photo 7.1

Photo 7.2

The casting, or block is first milled underneath, leaving the step on it. With hind sight I should have made the step much deeper than it is, for reasons that will be explained later. The hole is then marked out and drilled to a size that will give it a very sloppy fit on the tool post stud. Fix the mounting over the stud with the step hard against the side of the top slide then drill the hole for the spindle by pushing the saddle along the bed with the tail stock. Do not damage the tailstock barrel with the drill! Finish to size with a boring bar between centres. Drill and tap for the clamping bolts and finally slit through to the spindle hole. **Photo 7.3** shows the spindle parallel, then with the spindle reversed in the mounting, at right angles, **Photo 7.4**. Any other angular drilling is taken care of with the top slide protractor.

I had the occasion to make a six spoke hand wheel for a model, **Photo 7.5**, and this was made using the spindle. The next series of photographs show the procedure for making a wheel, this time with four spokes. A blank of the required diameter was prepared and was held by the outside of the rim in a collet in the lathe chuck. **Photo 7.6**. A drill was selected to give the radius required at the ends of the spokes. Packing equal to half the diameter of

Photo 7.3

Photo 7.4

Photo 7.5: Caption

Photo 7.6

Photo 7.7

the drill, plus half the intended width of the spoke was placed under the spindle mount before it was tightened down. It is for this reason that a deep step is to be preferred. The back was again held hard against the side of the top slide. The chosen drill was then moved to a position just inside the rim of the wheel nearest to the operator and the cross slide locked. The tool described in chapter 6 was used here. My lathe is fitted with one of the common methods of dividing, i.e. 24 holes drilled around the chuck back plate and with a detent attached to the headstock. Four holes were then drilled around the blank, starting each with a centre drill. **Photo 7.7.** Next the drill was moved to the same distance to the opposite side of the lathe centre line, again using the offsetting device, and four more holes were drilled using the same holes of the dividing method. There were now eight holes representing the outside ends of the four spokes. **Photo 7.8.** All of this is now repeated for the inside ends. **Photo**

Photo 7.8

Photo 7.9

Photo 7.10

Photo 7.11

Photo 7.12

Photo 7.13

Photo 7.14

7.9. At this point the wheel in effect has been marked out with precision and could be removed from the lathe to be sawn out and hand finished in the conventional way. However things can be taken a stage further. If a small slot drill is substituted for the twist drill, then the inside of the rim and the outside of the hub can be milled out by slowly turning the lathe mandrel by hand against the rotation of the cutter. If the dividing arrangements on the lathe are of the worm type then, providing the cut is a very light one, this could be used to rotate the mandrel. **Photos 7.10** and **7.11**. The sides of the spokes are then cut as in **Photo 7.12**, and finally the blank is ready to be removed from the collet to be hand finished, although the amount required has been reduced to a minimum. **Photo 7.13**.

Although this method only makes parallel spokes, or crossings as clock makers would say, if, when hand finishing, each spoke is filed with a radius that is greater near the rim than the hub on each side and front and back, then an illusion of taper can be produced. Clamp the wheel horizontally with a spoke in space then thread a narrow strip of emery cloth over the spoke, and pull down on each end alternately. Soon all the facets are blended together to produce the desired effect.

May I, at this point, issue a safety warning from a lesson I learnt the hard way. Disconnect the power from the lathe motor, or by some other means make sure that the mandrel cannot be turned under power. It is so easy to forget and to switch the lathe on instead of the spindle. With my set up, a mishap causes rapid wear of the detent and the division holes, as well as making a terrible noise, but with the worm system it could cause a nasty accident to the operator and serious damage to the machine.

Some difficulty may be encountered when setting the spindle with the Off-set device. Because the packings under the spindle mounting block have lifted it above the rule, it is not easy to align it with the graduations. This is easily overcome. Remove the rule and replace it with a length of light angle, I use 3/8 x 3/8 x 1/16 aluminium. Re-attach the rule in a vertical position with small clamps, and align it with a centre in the tailstock as before. I have found that the rule graduations are tall enough to have covered all of my needs so far. Further to assist in the accuracy of the measurements, I use an ordinary sewing needle, with the eye cut off, gripped in the spindle chuck, to align with the graduations on the rule.

Photo 7.14 shows the arrangement.

This is not the end of the story of the drilling spindle, and we will meet up with it again in chapter 10.

PART TWO: FOR THE MILLING MACHINE

CHAPTER EIGHT SCREW SLOTTING

At some time the model engineer will need to put a slot in a screw head, or in the end of a rod. I have read instructions which tell them to use a hacksaw, even if necessary to put two blades in the frame to get the right width ! I can't think of a worse testimony to poor workmanship than a hand cut slot. It may seem that the time spent in setting up a machine to cut perhaps only one or two slots is not worth the effort, but with this fixture it can be done in only a few seconds.

Photo 8.1 shows my fixture, but many other variations are possible. The single dedicated T-bolt allows it to be rapidly mounted on the milling table and aligned with a square, or it can be gripped in a machine vice.

The first thing to do is to find a suitable chuck, in my case a very old, but still serviceable lever scroll chuck. These can usually be found on the stalls of second hand machine dealers at shows or exhibitions. At a push a drill chuck could be used but they do tend to be rather long in the body, and have a small capacity for their bulk.

The body is a block of metal, steel, iron, brass or aluminium, whatever is to hand. Sizes are made to suit

Photo 8.1

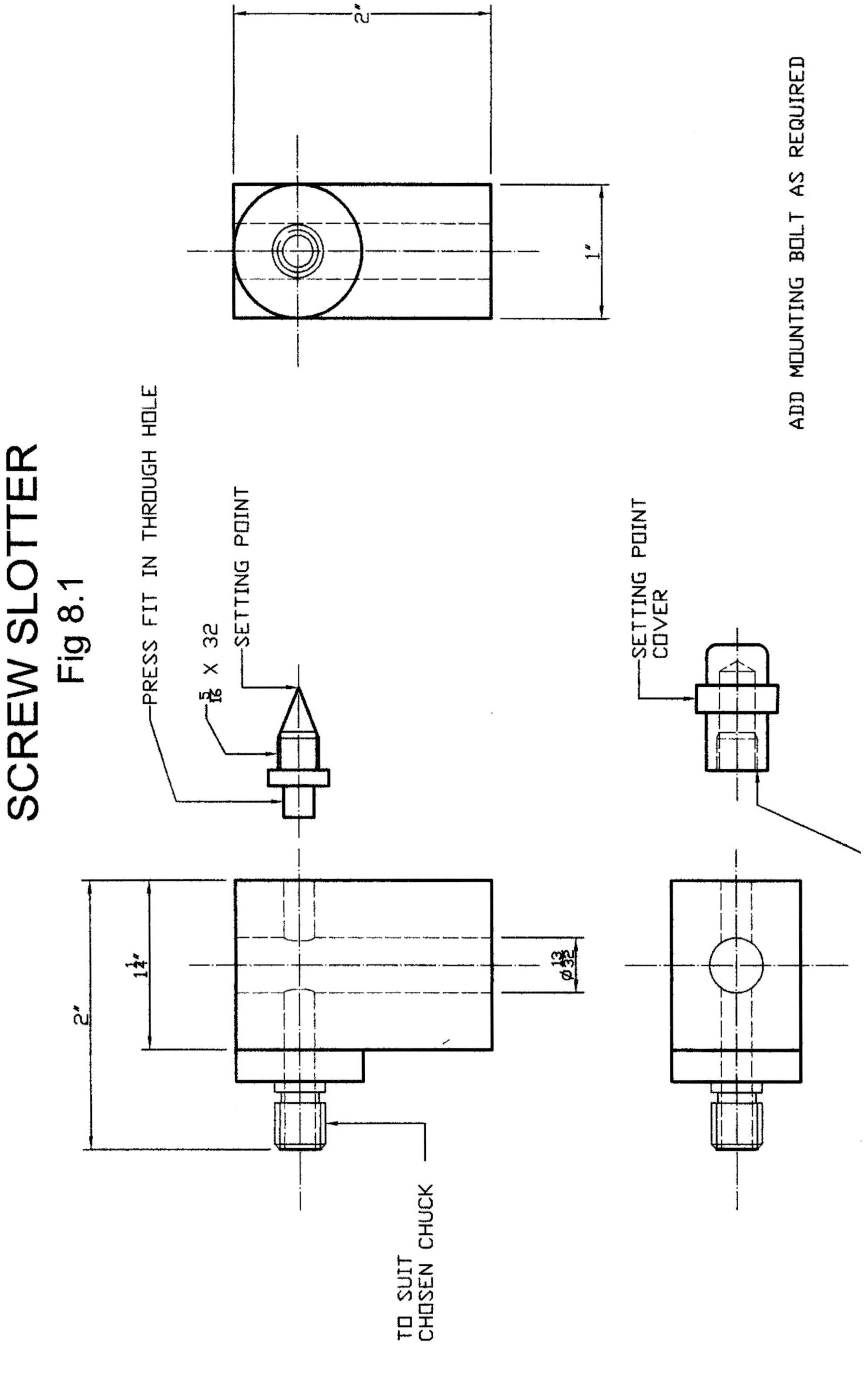
SCREW SLOTTER
Fig 8.1
PRESS FIT IN THROUGH HOLE
5/16 X 32
SETTING POINT
SETTING POINT COVER
ADD MOUNTING BOLT AS REQUIRED
TO SUIT CHOSEN CHUCK
2"
1 1/4"
Ø 13/32
2"
1"

Photo 8.2

Photo 8.3

your requirements, but as a guide mine started off as 2in. (50mm) high by 2in. (50mm) wide by 1in. (25) deep, **Fig 8.1**. In the lathe four-jaw chuck, face off one end square to the sides, then reposition and machine away the width to leave a spigot which can be threaded to take the selected slotting chuck. If a drill type chuck has been selected then drill and tap as required, but in both cases put a hole all the way through the block. Next drill a hole through the length of the body to take the holding down bolt. Please ignore the lower hole through the body, it happened to be already in the piece of scrap steel that I used. The final part is a little pointer that is pressed into the horizontal hole. Make a little screw-on cover for it, to prevent damage to your hands, or even worse, bend the pointer. A slitting saw is mounted on an arbor which can be screwed like a Clarkson cutter shank to fit a chuck, or a taper that can fit directly into the milling spindle, whichever is the quickest to set up. I find that a saw of one sixteenth of an inch thick covers practically all my needs.

To make a slot, the body is clamped to the milling machine table with the single bolt, a try square across the bed aligns it or it could alternatively be held in the milling vice. The saw is fitted into the spindle. Alignment of the saw in the vertical, Z, direction can be done in one of two ways.

1. A small pip is left on the component to be slotted and the saw is aligned with this, or,

2. the saw is moved to the back of the fixture and is aligned with the uncovered pointer. In either case, it only needs doing once per slotting session. The depth of cut is put on in the X direction and the length of the slot in the Y direction. A minor limitation with this design is the length of work that can be inserted into the chuck before being stopped by the central bolt, but in thirty years of use this has never been a problem for me.

By an amazing piece of good fortune (it will never happen again), when my chuck is tightened onto the fixture one of the jaws is exactly horizontal. I had the need at one time to make a batch of castle nuts from stock material and because of the position of the jaws, each of the hexagonal sides was perfectly vertical and the slot was right in the middle. Three slots were then cut all the way through and the job was quickly and accurately done. I think that this feature would prove useful if deliberately built in.

Photo 8.2, shows the few simple components, a grubby but accurate lever scroll chuck, the body, the mounting bolt and nut, and the pointer cover.

Photo 8.3, is a group of items made with the fixture.

Any poorly made piece of work is an eye-sore forever, while a good one is a joy for the same length of time.

CHAPTER NINE PAPER PLATES

There have been many different methods described over the years on how to divide particularly awkward numbers, many by using the holes in preformed strips, such as the edges of computer printouts, or the perforations on the edge of film. These are then formed into a ring with the correct number of divisions, and mounted on a disc to be fitted on the dividing head or lathe.

I would like to describe a method I devised some years ago to do the same. I have two dividing heads, one commercial and one home made. As with most model engineers, when cutting gears I have waited with trepidation to see if after one revolution, the cutter will pass cleanly through the gap between the first and second tooth. So far it always has done, but the stress I can well do without. My preferred method, therefore, is to use my home made head with the worm removed and the plate mounted on the end of the spindle for direct dividing. This of course means that in most cases a plate with the exact number of holes is required, but if one is not available this is not a problem as I can print one on a computer. Various examples can be seen in **Photo 9.1**.

I think that the most basic CAD programs (i.e. cheap) have the ability to do radial copying. To start making a plate, first draw vertical and horizontal centre lines. About the intersection draw a circle equal to the diameter of your normal division plate, mine is six inches. Again draw another circle, but this time slightly smaller than the mounting hole of the normal plate. Why smaller? It will be explained later. At a short distance inside the outer circle draw a much smaller circle, say about 0.125in. (3mm) diameter. This represents the hole. Using the radial copy/circular array facility, copy this hole as many times as is required. This is usually done by indicating the object, the number of copies required including the original, the angle of rotation, 360 degrees, and the centre of rotation. The computer doesn't care how many

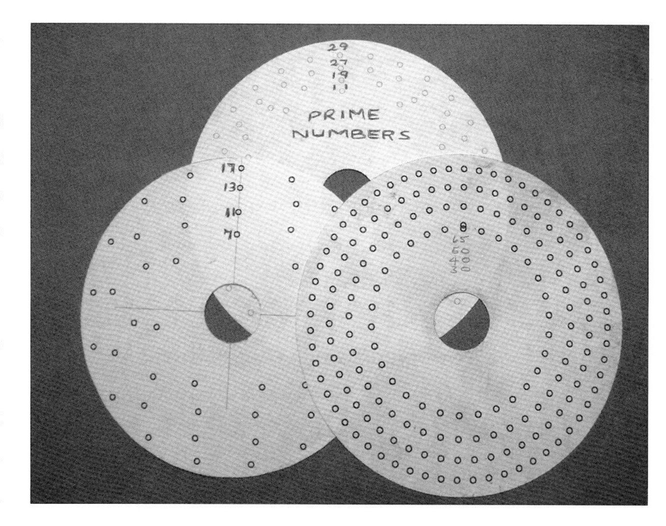

Photo 9.1

it has to draw, or if they are prime numbers (Note:- *CAD for Model Engineers* by D.A.G. Brown, Workshop Series No. 29, explains this type of copying). Personally I then delete the centre lines but this does not matter one way or the other. The 'plate' can now be printed out using the printer's maximum dots per inch. Thick paper or thin card is best.

Roughly cut round the outside with scissors then place, printed side down, onto an ordinary division plate and adjust until they are truly concentric then clamp with strong paper clips of the bulldog type. The 'slightly smaller' circle should be clearly visible in the centre hole of the metal plate, as a check of concentricity, as this is very important. With a very sharp knife, of the scalpel type, cut out the centre of the paper or card, using the metal as a guide. Trim up the outside, write on any information that is required, and the paper plate is finished.

The new paper plate is fitted onto the head spindle being backed by an ordinary plate and clamped with a large washer. **Photo 9.2**. As can be seen in **Photo 9.3**. I rather

Photo 9.2

Photo 9.3

went to town over the detent, just for the fun of it, (but isn't that what it's all about?), a much simpler reference point would do just as well. All that is required is some kind of fixture with which to align the printed holes.

The top knob allows the fiducial to be slid along its slot to cover different rings. The fiducial itself is a piece of shim steel which is bent to press lightly on the paper, so in effect giving a knife edge to align the circles with. It should also be truly radial to the spindle. I always choose to align the bottom edge of the circles. The lower knob operates what is a kind of calliper brake. Nylon pads clamp the plates front and back and are pivoted vertically so as to be self aligning, so preventing the plates from turning. This is used as well as the usual spindle lock. The bigger the hole circle is the better the angular accuracy.

Some years ago, the system was exhibited at Alexandra Palace where it received some criticism as to accuracy, being regarded as only suitable for items like cylinder covers. I would certainly argue about that, and though I would never claim this to be a precision system, I do believe it is as good a way of getting practical results for unusual and prime numbers as many others, and better than some. Cards, once printed, can be kept and used again if required. I have used them to produce several sets of gears for unusual ratios, all of which run smoothly together.

POSTSCRIPT

After writing this chapter, I wondered if there was any way it could be used with my other dividing head. This is a Vertex B.S.0 which I believe is quite common amongst model engineers. An article in a magazine some years ago indicated the need to make an item with 63 divisions, so I checked the tables in the manual and found that this was not possible with this machine and the division plates supplied, as compounding was required and would need the more complex B.S.2 Head. However I thought that it might be possible to make a paper plate to give the necessary divisions and after some experimenting it was so proved.

The angle between two adjacent 63 divisions is 5.71428 deg.

The B.S.0 ratio is 40:1, therefore one turn of the handle is equal to 9 deg.

If a paper plate with 63 holes is made then 1 hole equals .14285 degrees . (9/63).

If 40 of those 63 holes are counted off then 0.14285 x 40 equals 5.71428 deg. - the target division.

A plate with 63 divisions was quickly made up and attached to one of the commercial plates supplied with the machine. The fingers were set to count 40 divisions, and the zero on the output face plate protractor lined up with the marker on top of the head. Sixty three movements later the zero was again in line, so obviously it worked.

However there are a few points that need watching.

1. The paper needs backing up with a normal plate but the circle of printed holes must lie between two rings, or the detent may punch through and cause misalignment.

Photo 9.3

Perhaps a better solution would be to make up a dummy backing plate.

2. The outside of the plate is 100mm in diameter and the bore is 21mm. I found that a printed circle of 88mm dia. was suitable. The printed holes should be about 3mm in diameter.

3. Cut out the paper as already described for the direct head and stick it on using small pieces of double sided tape (carpet tape is ideal). Cut holes for the backing plate mounting screws.

4. On the B.S.0 the fingers are pressed on to the plate by a spring clip fitting into a groove. The addition of the thickness of the paper makes this tight, but with care it can be managed.

5. Because the detent is not positively located, care must be taken when moving the fingers so as not to knock it.

6. **Photo 9.3** shows the set up.

7. This then is another way of using paper plates. Personally I would only use it as a one off. If I wanted to make more than one item I would use the paper plate to make a metal one with drilled holes. Any slight radial error which may possibly occur with the paper plate will be reduced by 40 times on a drilled one and then again by 40 times on the workpiece. I do not know how far this can be used as a substitute for compound dividing and I do not intend to try and find out, but I do offer this idea as a possible way of getting round a workshop problem.

CHAPTER TEN SIMPLE ENGRAVING

If your tool post drilling spindle is of the independently powered type as shown in Chapters 6 and 7, then it could be adapted for another use, although it may be limited to a particular type of milling machine. In my case the machine in question is a Centec 2B, a medium sized machine for the home workshop, with a fixed head and a movable knee. Milling or engraving with very small cutters requires a high speed, sometimes above what the machine is capable of. With the simple fixture shown in **Photo 10.1** the drilling spindle is held parallel to the main spindle. It must have a stay to prevent rotation, which could prove difficult if the machine is of the fixed table type and where the cut is put on with a movable quill. However, with a suitable machine the setup has three parts:

Photo 10.1

1. A shank to hold the frame, 2. the frame, and 3. a stay at the back. The shank has a taper to fit the milling machine spindle, held by a draw bar, and a parallel portion which fits into one end of the bracket, secured by a couple of grub screws at the back. The shank shown in **Photo 10.2**, also doubles as a long reach slitting saw arbor,

Photo 10.2

Photo 10.3

Photo 10.4

hence the threaded portion at the bottom end. This serves no purpose in the present set-up. Also ignore for the time being the block with four cap headed screws, fixed to the large end, its purpose will be explained later. My bracket is a simple casting in aluminium, but it could be fabricated from other materials. The hole to hold the drill spindle must be bored parallel to the shank hole, and then slit so that two clamping screws can grip it. The stay is a rigid strap that fastens to the main body to prevent rotation of the mill spindle. **Photo 10.3** shows a view from the back.

VERY IMPORTANT – before setting up, disconnect the mill from the power source, to prevent accidental starting.

With the workpiece fastened to the table and using very small cutters, either bought or home made, simple engraving can be done by means of the feed screws. A rotary table can also be employed. The Roman numerals on the clock shown in chapter 14 were done with this setup.

Now to explain the block on the end of the bracket. It is a tool holder. I had to machine some V-grooves in the table of the Nut Cutting machine (also shown in chapter 14), and not having a suitable horizontal, or vertical cutter, I attached the block to hold a square tool ground to the correct angle. The Centec 2B has a rapid traverse handle so the grooves were rapidly planed into the model, **Photo 10.4**.

CHAPTER ELEVEN A SAW BLADE SCARFER

One of the most useful pieces of equipment in my workshop is what is usually referred to as a vertical/ horizontal band saw. Many projects which involve heavy cutting would never have got started, due to advancing years and a heart condition, without it. For economical reasons, blades are made up as required, from a 100 foot roll, cut to length, chamfered then joined by silver soldering. I have never really mastered the art of easily filing the chamfer on the two ends. Trying to tighten a vice while holding the blade at the correct height and angle within soft jaws, and six feet of blade with a mind of its own, thrashing about on the floor, is something I very rarely get right first time.

With this in mind, I made this simple fixture. **Photo 11.1**. It is in effect another filing rest, but as it is not for use on the lathe as the rests in Chapter 1, I decided to give it a chapter of its own.

There aren't many components, just two side plates, two pivots, two rollers, two distance pieces, two compression springs and four 6BA counter-sunk screws. **Photo 11.3**.

Cut the side plates from 35 x 6 mm. flat plate or material and mark out the holes as per drawing, **Fig 11.1**. Scribe the line for the centre of the V-groove. Clamp both together and drill, the top holes are to just clear 8BA, while the bottom two are to clear 6BA, or their equivalents. Fasten both sides together with a couple of bolts in the top holes.

To mill the grooves, set the plates in the vice with the scribed line parallel to the jaws. I often use pieces of lathe tool steel as parallels, and it simplifies setting if a piece is first clamped into position whilst the plates are held flat in the bench vice. They then are merely dropped into the mill vice with the parallel on top of the jaw, before tightening the vice, and removing the clamp. A fly cutter, ground to an angle of 90 deg. is then used to cut a groove of 1.27 mm. (.050") deep. Without removing the plates from the vice, take the cutter round the back and make another groove to the same depth. **Photo 11.2**.

The bottom holes in each plate are now deeply countersunk for the 6BA screws on the opposite side to the grooves.

Photo 11.1

Photo 11.2

SAW BLADE SCARFER

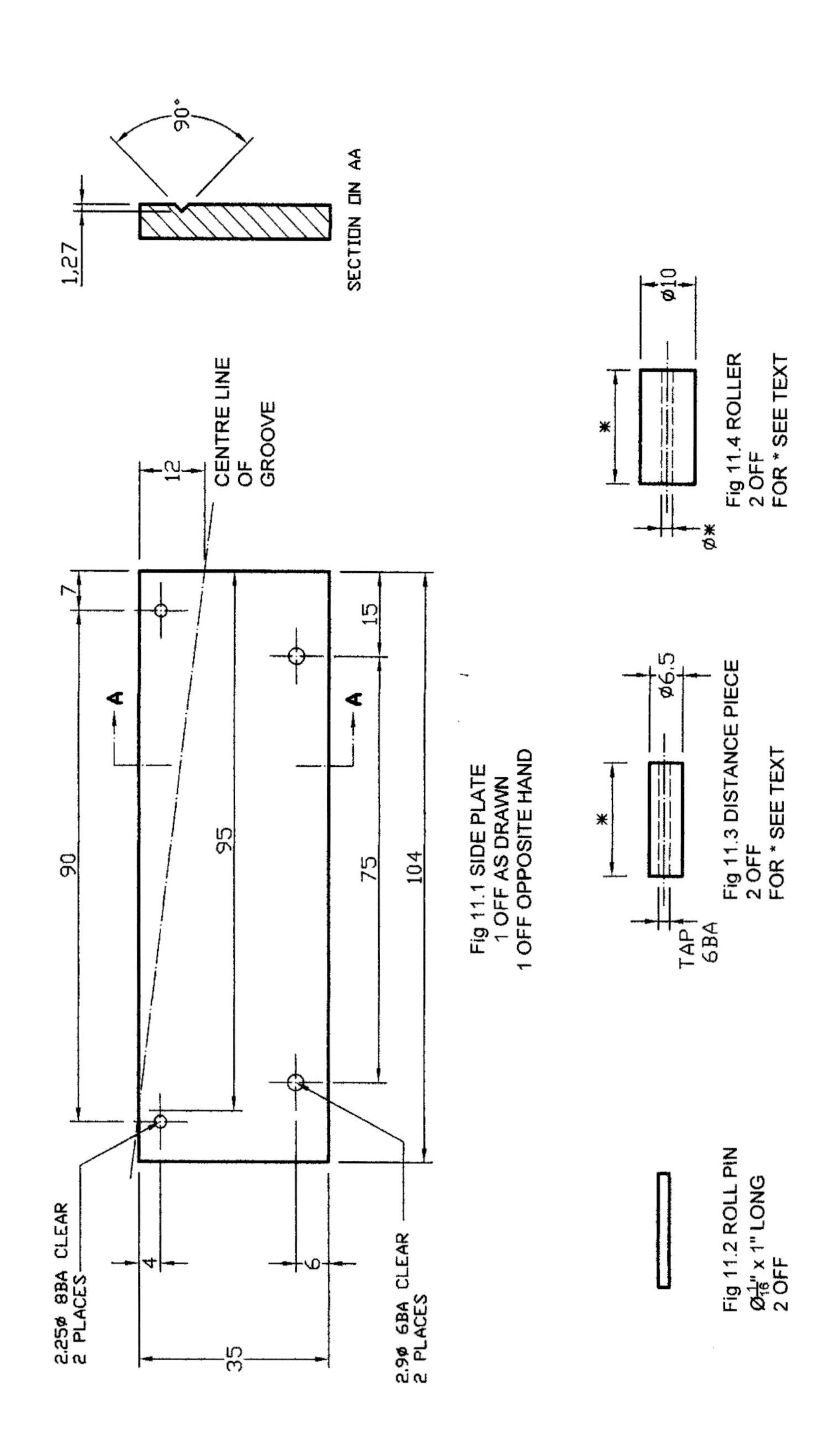

Fig 11.1 SIDE PLATE
1 OFF AS DRAWN
1 OFF OPPOSITE HAND

Fig 11.2 ROLL PIN
Ø$\frac{1}{16}$" x 1" LONG
2 OFF

Fig 11.3 DISTANCE PIECE
2 OFF
FOR * SEE TEXT

Fig 11.4 ROLLER
2 OFF
FOR * SEE TEXT

2 OFF LIGHT COMPRESSION SPRINGS (TO FIT OVER DISTANCE PIECES)
4 OFF 6BA COUNTERSUNK SCREWS

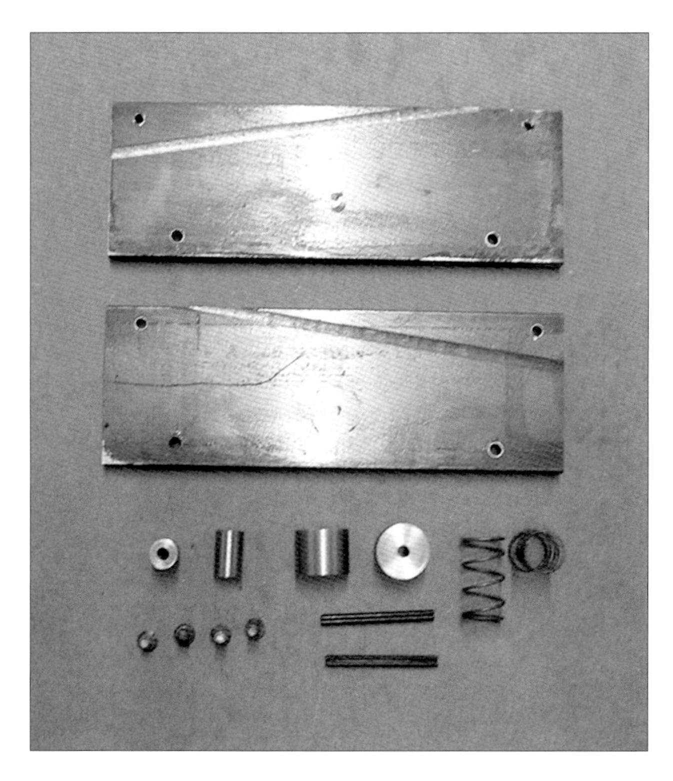

Photo 11.3

The top two holes are for the pivots of the rollers, **Fig 11.2**. I chose to use spring roll pins 1/16 x 1in. long for these because of their wearing properties, but this is not absolutely necessary. Plain round steel pins glued in will do just as well, but I would recommend using silver steel. Open out the holes to suit the method you have chosen.

Next the rollers and distance pieces have to be made to the lengths measured from the job. A piece of new band saw blade about 100 mm. (4in.) long is required. This is placed in the grooves and the distance between the side plates is measured. To do this whilst holding it all in the hands is well nigh impossible, but there is a simple way of doing it. Take a couple of screws about 25mm. (1in.) long and put them through the counter-sunk holes and secure them with nuts on the grooved side. Put another nut on each and run them down until they are about equal to the width of the blade. Put the second side on with the sample blade in the grooves and secure with a rubber band around the middle. Each corner can now be measured and the loose nuts moved up or down the screws until all are equal. This is the dimension used to work out the lengths of the rollers and the distance pieces. The distance pieces are 6.5mm. (0.25in.) diameter tapped 6BA both ends and the length equal to the distance between the plates, less 0.2 mm. (.010in.), **Fig 11.3**. The rollers are 10mm. (0.473in.) diameter, and the length less .64mm (0.025in.), with the centre holes drilled to give a running fit on the pivots, **Fig 11.4**.

To assemble the parts, take one of the side plates, it doesn't matter which, open out the roller pivot holes and put the pivots in from the groove side, leaving 15mm. for the rollers. Do not cut off that which protrudes from the other side, it's needed. On the same plate fasten the two distance pieces with countersunk screws. Before the second side can be fitted, the top holes that support the outer ends of pivots need to be opened up to be a loose fit over them. The idea is that when the scarfer is put in the bench vice, this side plate will rock against the distance pieces and grip the saw blade only, even allowing for slightly different widths due to different tooth pitches. Slide a couple of light compression springs over the distance pieces and fit with countersunk screws, but do not tighten, leave them loose by about a couple of threads.

The device is now ready for use. Put it loosely in the bench vice, the bits of the pivots that stick out now rest on the top of the jaws and hold it in position. Insert the length of saw blade into the V-grooves so that the square cut end rests just on top of the centre of the roller. The positioning of the blade is important as it prevents the end from bending away under the action of the file when it gets very thin. Gently nip up the vice. File away until no more metal is being removed, this can be done in either direction. **Photo 11.4** shows the result on a short test piece of blade. Remember when filing the second end, to put the blank into the scarfer with the blade teeth on the same side as the first end. Within reason, the longer the scarf is, the more reliable the joint is likely to be.

Photo 11.4

Chapter Twelve DIVIDING ARMS

As already mentioned in chapter 9, I possess one of the popular Vertex BS0 dividing heads. Although very useful it does have a fault which can cause problems when multiple turns of the indexing handle are required. The detent, when it is pulled out of the index holes for turning, cannot be latched and the spring, which is quite strong, tends to make it creep back in until it catches the indexing fingers, moving them from the correct position and so spoiling the count. Disaster.

To correct this manufacturer's oversight and fitting a latch takes only a few minutes.

Disconnect the arm from the machine. After marking the position of the end of the plunger in the outer sleeve, knock out the roll pin with a punch. Dismantle the outer sleeve, spring and plunger, but there is no need to remove the inner sleeve from the arm. Decide on a piece of rod for the peg, mine was an oddment of 0.110in. diameter, about 2.8 mm. Drill a suitable hole through the inner sleeve at 11/16in. (17.5mm) from the face of the slotted arm. It can be a forced fit for the peg or fixed with adhesive, but it should not protrude on the inside where the spring slides. Cut it off at about 3/32in. (2.4mm) high. In the outer sleeve mill a slot in the open end 9/16in. (15mm) long, and of a suitable width to easily slide over the peg. Turn the sleeve about one eighth of a turn and make another shorter slot, the length equal to about half the diameter of the peg. Reassemble the components with the roll pin and the job is done. **Photo**

Photo 12.1

Photo 12.2

Photo 12.3

Photo 12.4

Photo 12.5

12.1 shows the detent in the latched position, well clear of the indexing fingers.

Another feature to consider if you are making your own direct dividing head, or adapting the lathe for direct dividing, is to make the detent arm adjustable for position. **Photo 12.2** shows a home made dividing head and **Photo 12.3** is the idea applied to the lathe. Occasionally it is necessary to relate a series of divisions to a location on an existing surface, an example of which would be the drilling of holes in the flat sides of a hexagon. It would be extremely lucky if one of the required holes in the division plate lines up exactly with the flat on the hexagon. **Photo 12.4** illustrates this. With the modifications suggested it would be a simple matter to loosen the detent arm and rotate it until the flat is in the correct position before retightening. **Photo 12.5** shows the workpiece in the corrected position.

Chapter Thirteen

A GRADUATING TURRET

The final items are two variations of a simple theme, and will greatly assist in the making of graduations. The first is the common ram type of engraver that is usually employed on lathes, and the second, using the same principle, for guiding the stylus on a Taylor-Hobson engraving machine, shown in **Photo 13.1**.

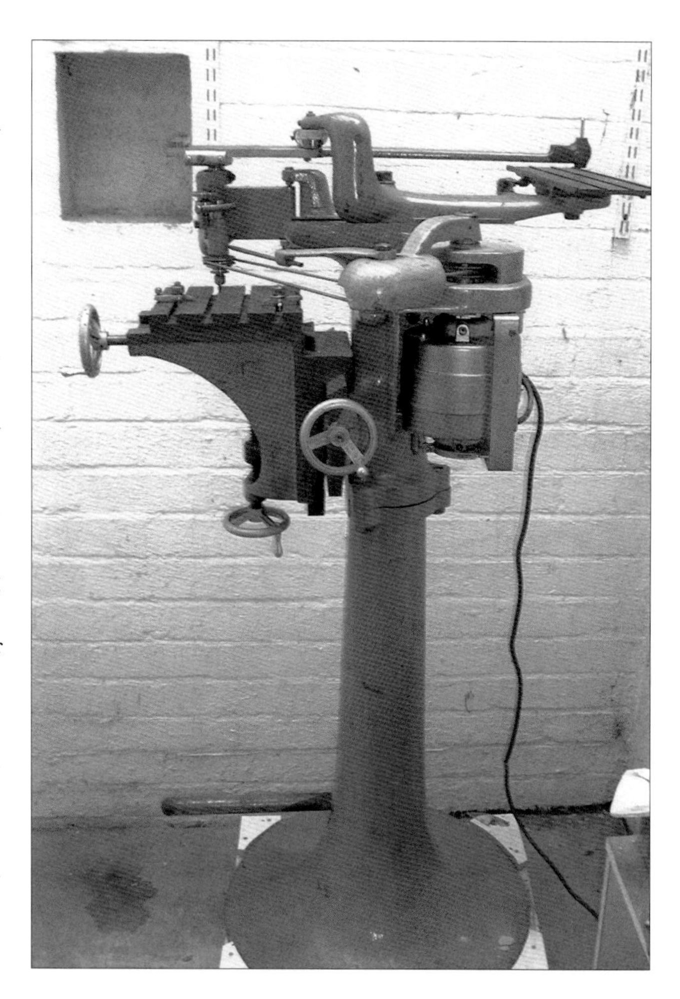

Photo 13.1: Taylor-Hobson Engraving Machine

To make a scale, either linear or circular, lines of equal spaces and various lengths have to be engraved. The spacing is taken care of by arranging the workpiece to move in a linear way, or by a fraction of a circle. To allow the finished scale to be easily read, some of the lines are given prominence by making them different lengths, such as the fives and tens on a protractor, or eighths, quarters and halves on a rule. This is usually done incorporating a form of turret, usually vertical, into which are screwed adjustable stops which limit the amount of travel of the ram. That only leaves the problem of counting the number of lines of one length before changing to one of a different length. For example, on a protractor scale marked in degrees there would be a long line to start with, then four short ones, then a medium one to indicate five, then four more short ones and finally another long one for the ten, before starting all over again. Several suggestions to aid this counting have been made over the years, such as using pencil and paper, or little piles of counters (washers) which are moved as appropriate as each line is made. Doing it this way it is very easy to make a mistake – I know from experience! With my simple horizontal turret, both the length variations, and the counting are taken care of. All that is necessary is to turn the turret after each line is made.

It may be that existing ram type engravers, or the designs for them, could be adapted to this turret system, however if the reader intends to start from scratch and make their own, then here are a few of the features I built into mine, which is to be used on the PL1018. Sizes would need to be made to suit the intended lathe, but a general idea of the proportions can be seen in the various photographs, the main block is 3 x 2.5 x 2in, (76 x 70 x50mm).

The item shown in **Photo 13.2**, is the ram type, while **Photo 13.3**, is the Taylor-Hobson version.

The turret is very similar for both applications being a disc, in my case aluminium alloy, about 1/8in. thick. It does not call for any great precision and can be marked out by hand using dividers and a rule. From the centre, scribe a series of circles which correspond to the various lengths of the lines, then from the centre, radial lines corresponding to the number of divisions required. No great accuracy is needed for the radial spacing. At the appropriate intersections, centre pop and drill the holes, then saw and file the slots. The width of the slots should give plenty of clearance for the stylus of the T-H, or the stop on the ram. The centre hole should be free fit on the pivot bolt with no sloppiness. A wavy washer under the hexagonal head gives a slight turning resistance to the disc.

It will be observed that the difference in the lengths of the slots in the T-H disc is rather pronounced. This is due to the way the engraving machine works. The cutter head is fixed to the end of a pantograph arm, which, even at its lowest setting reduces everything by a factor of three. Therefore the difference in the length of the slots in the template (discs) need to be three times longer to

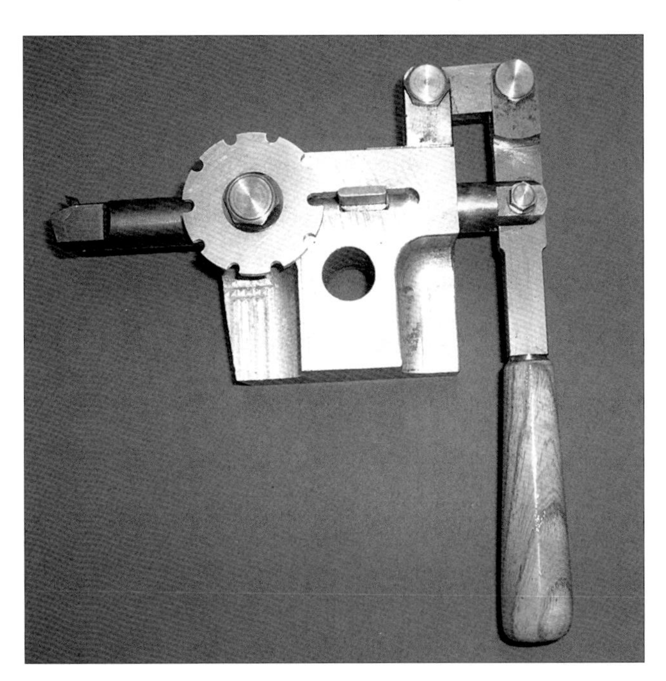

Photo 13.2

Photo 13.3

Photo 13.4

Photo 13.5

compensate for this. A ram type engraver does not have this problem so the discs can be made much smaller than the 2.125in. diameter seen in the photograph. As with all these length of line devices, they only control the variations of the lengths, and not the overall lengths of the lines. This length is set by the positioning of the ram to the edge of the workpiece.

Let us start with the body block. This is from, once again, a convenient lump of aluminium alloy, which is just about my favourite material. The underside is machined flat but has a step left on it just as on the tool post drilling spindle in chapter 7, **Photo 13.4**. A mounting hole is drilled through it, and also another hole, whose position and purpose will be explained later. The reason for the step is as follows. Often the dial being made needs to be bevelled where it is to be engraved, and this is machined with the top-slide set to an angle. If the ram is then placed on the top slide with the step hard up against the side, without altering the angle of the top slide, then the travel of the ram will exactly match the bevel on the dial with no extra fiddly setting up.

Photo 13.6

After making sure that the top-slide is parallel to the lathe centre line, drill, bore and ream the hole for the ram, which will now be at centre height. I chose 0.625in. diameter for the ram, because I had a reamer of this size, and some bright mild steel for the ram. A keyway 0.250in. wide needs to be cut exactly above the ram hole, and a tapped hole for the turret pivot. The extra hole, underneath is drilled (0.25in.- 6mm) through into the ram hole at about the middle of the length of the key slot.

Next, the ram. There are three bits of machining to do on this component, the hole for the cutter, the slot for the handle and the key seat. The first two must be in line with each other and the key seat at a right angle to them. Mill the slot for the key 0.625in. long, but before removing the ram from the vice, drill a couple of holes, 6BA clear, from the bottom of the slot, right through. Now remove from the vice, turn over and counter-bore these holes to take the head of a 6BA cheese or socket screw.

The mild steel key is made from stock to be a good fit in the seat, but is left very tall. The part that sticks above the guide slot is reduced to about 0.125 in. thick, so that the slots in the turret do not have to be too wide. Using the ram as a jig, drill and tap two 6BA holes for fixing. The extra hole in the bottom is used to insert the key when fixing the screws on assembly.

I have chosen a horizontal handle and linkage to drive the ram because of its simplicity but other systems could be used, **Photo 13.5**.

Cut away the block to clear the stroke of the handle, and also for aesthetic reasons, if so desired.

A turret was made as already described, for ten divisions, as it will only be used for circular dials where degrees are counted in tens. A long line is for tens, a medium for fives, and a short one for those in-between. Of course a turret with any other variation can easily be fitted. It may be noticed that the ram key is much taller than it really needs to be. This is so that some time in the future, perhaps, a system can be devised that will turn the turret automatically, operating from this key. But this would be a complication which for the moment is against the object of 'keeping it simple'.

Photo 13.6, set up and ready to go.

The T-H device is much simpler as can be seen in photo 13.3. A shallow groove guides the pantograph stylus into the turret. The two knobs on either side are part of the means of fixing the device to the dovetails in the table. It is shown fitted to the table in **Photo 13.7**. Again a turret of ten lines is used, for protractors and metric scales, but there is also an alternative one for making Imperial linear scales. The two turrets are shown in **Photo 13.8**.

Photo 13.7

Photo 13.8

GRADUATING TURRET

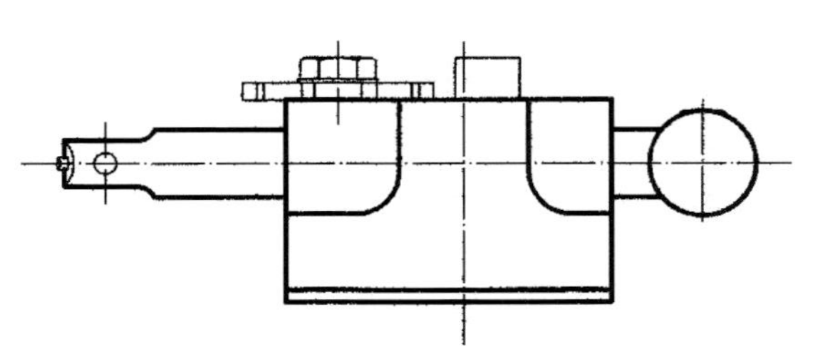

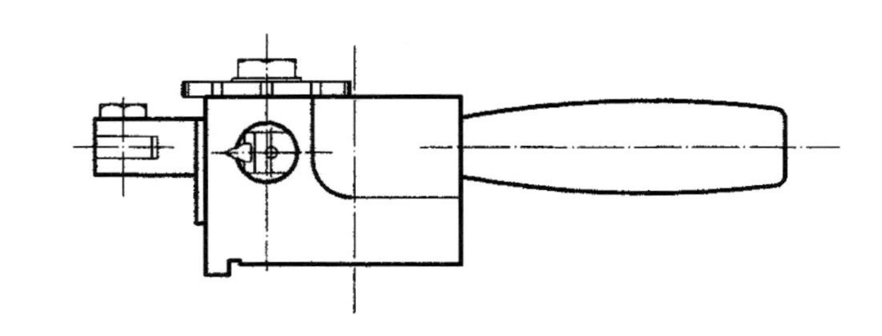

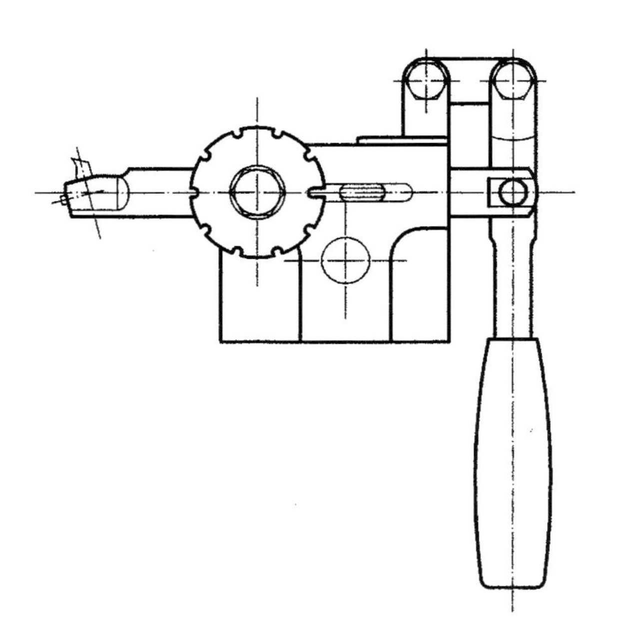

CONCEPT
GENERAL ARRANGEMENT DRAWING

Chapter Fourteen PHOTOGRAPHS OF MODELS MADE USING THE ABOVE EQUIPMENT

All the devices or procedures in the previous chapters have been used over the years to make modelling easier or to improve accuracy. For this final chapter I would like to describe some of the models I have made using them.

Photo 14.1 is a small vertical steam engine made from a design in a ***Model Engineer*** magazine from about the 1950's. All the castings were home made, aluminium for the base and the column, iron for the cylinder and brass for the flywheel. With the idea of making a batch of them, several of each of the castings were made, with the appropriate jigs and fixtures for machining them. However, as the only machine I had at the time was a treadle operated lathe, c1890, only the first one was ever finished.

Photo 14.2 is the Trevithick engine after the design by Edgar T. Westbury. The flywheel at 9in. diameter is the largest brass casting I have ever made. Over the years it has been on show in many exhibitions and has chugged away on low pressure compressed air for day after day after day. At home it has been steamed by placing it over a gas ring, to the amusement of visitors.

In **Photo 14.3** is a collection of three internal combustion aircraft engines. On the left is the Matador, a 10cc. OHV engine from a design in a magazine. In the centre is a twin cylinder, side valve, boxer type engine to my own design, with a capacity of 20cc. On the right is the Jones .605 two stroke engine of 10cc. capacity with glow plug ignition. This used the only casting I have ever bought. Whilst all of them would run I never really trusted their reliability and so none was ever installed in the radio controlled aircraft they were intended for. I must add that in the case of the Matador and the Jones this was not in any way due to the design but in my manufacturing.

The next series of photographs are three of an intended quartet of 19th. century machine tools, all made to the same scale of 1:12. **Photo 14.4** is a model of a Nut Cutting machine made by the firm of Archibald Milnes of Glasgow. This was used in the period before the ability to roll hexagonal bar stock, so that the large nuts, such as were used on the foundations of beam engines, were forged. After threading they were placed on this machine and all six sides were milled to fit a standard sized key, they didn't call them spanners in those days.

Photo 14.1: Vertical Steam Engine

Photo 14.2: Trevithick Engine

Next is a set of Angle Bending Rolls, **Photo 14.5**. These were often placed in a pit with the table at ground level to make the handling of long sections of angle easier. The three rolls could be turned over so that the bend could be made with the vertical leg of the angle on the inside or outside of the radius. Different forms of rollers could be used to roll other sections of steel. Three examples of rolled angles are also shown, an inside bend, an outside bend and one both ways. These were actually rolled on the model, though of course not in steel, they are lead.

What must have been an awesome machine in full size is the Cylinder Boring machine by Joshua Buckton & Co. of Leeds. This was capable of machining cylinder castings from 2ft. to 6ft. bore by up to about 10ft. long. A cutting head of the required diameter was fitted onto the boring bar. It had three cutters held in by wedges and three 'rubbers' which I imagine were some sort of hard wood to prevent chatter. The boring bar would be removed from its brackets before the casting was threaded over it, to be fastened onto the bed, and the bar replaced. It was rotated by the integral steam engine, and moved along the bar by the gear arrangement at the end, presumably from left to right. Two facing heads would machine the flanges, the tools being fed automatically by star wheels on the end of the lead screws. **Photo 14.6** is a general view and **Photo 14.7** is the steam engine. The head feed drive gears, which could be taken out of mesh for a rapid traverse by hand, are shown in **Photo 14.8**.

Photo 14.3: Three Internal Combustion Aero Engines

Photo 14.4: Archibald Milne's Nut Cutting Machine

Photo 14.5: Angle Bending Rolls

The fourth model is a Radial Arm Drill by Joseph Whitworth & Co. of Manchester, but is not yet far enough advanced to be included.

Two of the models, the Cylinder Borer and the Angle rolls were made from illustrations in a book called, *A Text Book of Mechanical Engineering*, published in 1900. The Nut Cutter and the yet to be finished Radial Arm Drill were from a book called, *The Engineers and Machinist's Assistant*, published in 1847. Working from old books brings its own problems because of the styles of drawing used in a different era. Often only two views are given so some shapes have to be guessed at and some times one view will contradict the other. Those books printed from copper plate etchings are even worse. Because the draughtsman, probably more of an artist than engineer, had to work in a mirror image, any right hand thread

Photo 14.6: Cylinder Boring Machine

Photo 14.7: Cylinder Boring Machine – the Steam Engine

Photo 14.8: Head Feed Drive Gears for the Cylinder Boring machine

drawn as seen would come out as a left hand thread when printed. This can cause confusion to the model maker as it is important. Common parts which were well known at the time were simply left off.

Photo 14.9 is a model of a Monastery Clock, again from a published design, but less the chiming mechanism because I didn't want it.

Finally **Photo 14.10** is 7 1/4 inch gauge model of the Londonderry Railway locomotive No.17. It was made by Head, Wrightson & Co. in 1873 at Thornaby-on-Tees. It is as yet unfinished. Measurements were taken from the original in a museum.

Photo 14.9: Monastery Clock